T0222380

CLASSICAL MECHANICS:
AN OVERVIEW

by

Richard Sparapany

authorHOUSE

AuthorHouse™
1663 Liberty Drive
Bloomington, IN 47403
www.authorhouse.com
Phone: 1 (800) 839-8640

Published by AuthorHouse 03/17/2016

ISBN: 978-1-5049-5383-2 (sc)
ISBN: 978-1-5049-5382-5 (e)

Print information available on the last page.

Any people depicted in stock imagery provided by Thinkstock are models, and such images are being used for illustrative purposes only. Certain stock imagery © Thinkstock.

This book is printed on acid-free paper.

Because of the dynamic nature of the Internet, any web addresses or links contained in this book may have changed since publication and may no longer be valid. The views expressed in this work are solely those of the author and do not necessarily reflect the views of the publisher, and the publisher hereby disclaims any responsibility for them.

in loving memory

of my parents

John Daniel

and

Vincenza Helena

CONTENTS

CHAPTER 0
PRELIMINARY CONSIDERATIONS

CHAPTER I
THE BASIC IMPLICATIONS OF NEWTON'S LAWS

Contents

CHAPTER II
THE EXTENSION OF NEWTON'S LAWS TO ALL
MATERIAL SYSTEMS

Contents

CHAPTER III
APPLICATIONS TO LINEAR MOTION

Contents

Contents

CHAPTER IV
THE ANGULAR PARAMETERS OF MOTION

Contents

CHAPTER V
APPLICATIONS OF THE ANGULAR PARAMETERS

Contents

CHAPTER VI
THE UNIVERSAL GRAVITATIONAL FIELD

Contents

Contents

PREFACE

This book is a presentation of an overview of Newtonian mechanics and is intended to impart to the reader a comprehensive understanding of its foundation and applications.

Regarding the first of these, Newton carried out a series of experiments designed to determine the motion of a mass in reaction to an applied force. Subsequently he abstracted from his experimental results, three simple laws which he claimed governed the motion of all masses. These laws are the foundation on which Newtonian mechanics is built.

In contrast to the simplicity of the three laws however, their application to the analyses of particular configurations of masses can quickly and easily lead to mathematical complexities and intricacies beyond the imaginings of those who have not previously encountered them. For example the detailed analysis of the movement of a liquid mass involves dividing the mass into differential chunks and tracking the change in shape of each of them as the motion progresses. The entire field of hydro-dynamics is an application of Newton's three laws to liquid masses. In the event that the mass is gaseous, its motion is further complicated by the volume, temperature and density changes that accompany its compressibility. These are the issues involved in the study of aero-dynamics.

These examples are the rule rather than the exception. Historically the methods devised to apply the laws in different contexts have pushed the frontier of applied mathematics ever forward thus establishing Newtonian mechanics as one of the main motivators to the advancement of this field of mathematics.

This scenario seems to have pervaded the applications of the three principles from the very beginning when Newton needed some of the elements of calculus to analyze his experiments. Calculus however had not yet been discovered. The issue was resolved when he devised the elements he needed for the occasion.

Note that between the generalities of the theory and the intricacies of their application, there stands the mathematics. It is usually addressed to the detailed description of the system under investigation whether it be the deformation of a beam under stress, the whirlpool in your sink, the explosion of a flammable gas, the collision of two galaxies or what-have-you. Implied by this remark is that the import of the theory and the scope of its applications is inextricably married to the associated mathematics whose simultaneous study considerably enhances one's appreciation for the magnitude of the laws. Hence as we delve into the theory's applications, the elementary aspects of the mathematics necessary to do so will be developed.

Although this book is neither a text nor a student course supplement it can be of some use in these contexts as an aid to the understanding of a college level introductory physics course. The principal contribution of the book however, and incidentally its more distinguishing characteristic is that the order in which the topics are considered and their style of presentation are conducive to a gestalt understanding of Newtonian mechanics.

The topics are arranged in an order that allows for an axiomatic approach to the subject matter using Newton's three laws as the axioms. This presentation conforms loosely to this idea: the three laws (axioms) are presented early on and are followed by an analysis of the most general conclusions that can be drawn from them. It then proceeds to a sequence of related problems which are considered by the author to be instructive and sometimes supplementary to the previous material.

The style adheres for the most part to that of an informal lecture whose content is sometimes the solution of a specific exemplary problem and other times, a discussion of an underlying philosophy. In either of these cases, there is often an allusion to a further development of the current issue which might be encountered in the event that a further study is undertaken.

It often happens that students and/or the intellectually curious have insufficient background in elementary calculus, vector algebra and vector calculus when they first encounter this subject matter. This difficulty compounded by the sheer magnitude of the subject leaves many of them with the feeling that they just missed something very mysterious and very exciting, and indeed there is some truth to this. All too often the sadness of this fact is intensified by the discouragement in which their frustration is encased – a discouragement which undermines the belief that they could ever partake of this wonderfully intellectual and emotional thing called physics, a gift given to us by some of the greatest minds in history.

The opening chapter on preliminaries is directed at the alleviation of this problem.

There is a twofold purpose not mentioned explicitly which guided the evolution of this presentation:

1. *to adopt Newton's three laws of classical mechanics as a complete world view of physical phenomena;*

> **2. to position classical mechanics in its relation to classical electro-dynamics, relativity and quantum mechanics, and to treat each topic in the context of its direct relation to Newton's laws, always with an eye toward its extrapolation and generalization to the many and diverse fields of their application.**

The manifestation of this underlying purpose is somewhat intractably laced throughout the final form contained herein.

In this context it should be noted that classical electro-dynamics takes exception to Newton's law of action-reaction: two charges moving in each other's electro-magnetic fields exert equal and opposite forces on each other in keeping with Newton's law, but the forces are not along their line of centers in contradiction to it. (These forces were unknown to Newton.) But this does not limit the range of application of Newton's laws: when the 'line-of-centers' assumption of the action-reaction law is eliminated in this particular context, Newton's laws are made applicable once again. In other words, this issue constitutes a correctable oversight.

In distinction to this consideration however, quantum mechanics and relativity represent real boundaries to the realm in which Newtonian mechanics can be applied: when we deal with dimensions of atomic size or smaller, things do not behave in a classical (Newtonian) manner and quantum mechanics must be used for the analysis; similarly, when we deal with velocities approaching the speed of light, things do not behave classically and relativistic mechanics must be used. Since these conditions are so extreme, it can be seen that the extent of physical phenomena left unaffected by these boundaries is enormous. It is this enormous bulk of physical phenomena that lends itself to analyses using Newton's laws.

For over two hundred years Newtonian mechanics dominated physics. Its realm of applicability knew no boundary until the advent of relativistic mechanics and quantum mechanics in the early 1900's. The boundaries imposed by these new theories were rather remote however: their effects became apparent only under the extreme conditions described above. But it is usually the case that elaborate apparatus is needed to observe phenomena under such extreme conditions: hence it is reasonable to say that

> ***Newton's theory governs all normally observable phenomena.***

Even today the understanding of his three laws and the mathematics developed to aid in their application provide a firm platform from which one can delve into the many and diversified issues they can address such as shooting a rocket to the moon or designing a bridge. Hence these laws and their applications are considered to be the birth of modern physics and engineering.

The implication of the preceeding paragraphs is that Newtonian mechanics provides a rather comprehensive world view. It offers a vantage point from which one can acquire an understanding of all observable physical phenomena.

In keeping with this last point, it is hoped that the reader, while wending his way through this material will experience an ever-growing appreciation for the stature of the laws of classical physics and for their extensive realm of applicability as well as for their simplicity, elegance and depth.

It was mentioned before that while the laws themselves have a certain conceptual simplicity about them, their application to real situations easily leads to elaborate and intricate mathematical issues. Whenever a current topic sufficiently motivates an excursion into a mathematical discussion, the excursion is taken. Hence the discussion of the basic implications of Newton's laws is preceded by a non-rigorous description of differentials and integrals. These are presented from the viewpoint of a physicist rather than from that of a mathematician.

Other mathematical excursions include discussions of the dot product and cross product of vectors, directional derivatives and integrals defining the flux across a closed surface. In some cases these things may be beyond the reader's ability to manipulate and evaluate mathematically, but nevertheless the symbols are introduced in a way that renders them descriptive of comprehensible physical processes in need of future quantization.

For the most part however, the book adheres to the basic principles of classical mechanics as taught in a first introductory course. It emphasizes the ever-presence of Newton's equation of dynamics and his law of action-reaction so that all the fragmented pieces of study usually encountered in a first mechanics course can be assembled into the single abstract concept called 'classical physics'.

There are a few exercises, several thought problems with solutions, analyses and commentaries. However, the pluricity of problems to exploit the details of each topic, the definition of units and the subject of dimensional analysis are almost completely neglected as they would serve only to infuse many pages into the book and thus dilute the purpose for which it is written. Such things abound in the textbooks where they are well represented and accessible to everyone. They are replaced here with a number of representative problems which are worked out in the text and are accompanied by discussions and analyses directed at understanding the underlying physics rather than developing a prowess in problem-solving.

If the reader emerges from these pages with little more than the knowledge that a system of mechanics is a rather comprehensive world view of the universe and the feeling that he can comprehend at least some of it, then in my eyes, the book is successful. If in addition, he can glean some of the methodology and details of analysis, so much the better. In any case no matter what your position with respect to these things, enjoy learning about the universe in which you live.   RS

Preface

INTRODUCTION

Physics is the study of the universe. Its subjects include everything from sub-atomic particles to the entire cosmos and they subsume such diversified subjects as sound, light, the motion of material bodies as well as their deformation, atomic structures, astronomy, electromagnetism, hydro- and aero-dynamics and countless others. In the study of any particular topic, measurements and experimental results produce an explanation of the associated phenomena. They are usually expressed in terms of a collection of tabulated results or a few simple relations.

A *theory* is a set of principles or equations which explains the items in such a collection for a large number of different experiments. It is therefore the abstraction from the experimental data of a general rule which seems to govern the behavior of the objects in all those experiments. Note that a theory is neither true nor false: it is just a theory. It is hoped however that there are more experiments, (not yet carried out) whose results (when they are obtained) also seem to be explained (governed) by the theory. Then these experiments are said to uphold or support the theory in question.

The theory is then tentatively held as a description of how things behave and is used to predict the results of future experiments. A result which contradicts one such prediction brings the theory's validity into question. The theory must then be refined to cover the current result or forsaken to be replaced by one which does. In either case the new theory must of course explain all the previous results as well. In this way, physics attempts to abstract the most general principles which explain all physical phenomena.

The first such theory came to us from Sir Isaac Newton who proposed the following three principles. asserting that they govern the motion of all material bodies:

1. **LAW OF INERTIA:** *in the absence of an external force, a body will move with constant speed (which may be zero) without changing direction;*

2. **LAW OF DYNAMICS:** *the acceleration of a body is directly proportional to the force acting on it and inversely proportional to its mass*

$$\vec{F} \ = \ m\vec{a} \ ;$$

3. **LAW OF ACTION AND REACTION:** *if a body A exerts a force on body B, then body B exerts an equal and opposite force on body A; these two forces are directed along their line of centers.*

A few remarks are in order here.

First, it should be noted that *force* is a basic notion familiar to us all in a natural way: it is any push or pull and cannot be defined in simpler terms. It has two properties: magnitude and direction: an object can be pushed hard or easy and we therefore speak of the 'size' or 'magnitude' of a force which is a very definite quantifiable attribute; also when a body is pushed it is pushed in a specific direction. Vectors, which are mathematical objects having magnitude and direction were devised specifically to describe forces and many other quantities pertaining to physics.

Second, there are only four types of forces known at present: **contact forces** (when the bodies are touching); **electrical forces** (like charges repel, opposite charges attract); **gravitational forces** (every two masses attract each other like the sun and the earth); **nulear forces** (strong, short range forces in the vicinity of atomic nuclei). The last three are not contact forces as they act across a distance.

Third, Newton actually expressed his second law in a different form which incidentally is more correct; he wrote it as

$$\vec{F} \ = \ \frac{d(m\vec{v})}{dt}$$

recognizing that it was the momentum ($m\vec{v}$) that was more directly related to the force. As long as the mass m of the object in question is constant, this is equivalent to the more usual form $\vec{F} = m\vec{a}$. An important exception to this is the description of rocket propulsion where the mass changes during flight by the ejection of fuel. In this case, $\vec{F} = m\,\vec{a}$ will produce incorrect results; Newton's original form of the second law must be used.[1]

Fourth, the first law, the law of inertia is superfluous since it is an immediate consequence of the second: if the external force is zero, then

---

[1] It is interesting to note that calculus was unknown before Newton. He devised it to describe the motion of bodies. Calculus was also discovered simultaneously and independently by a mathematician by the name of Leibnitz. Before these discoveries, it was noted that bodies acquired a velocity when acted on by a force. As a result it was believed that force and velocity were directly related. Newton's measurements of course did not bear this out: it was found that force was related not to the velocity but to the rate of change of the velocity which is now called the acceleration.

$$\vec{F} = \frac{d(m\vec{v})}{dt} = 0.$$

Thus the momentum $m\vec{v}$ and hence the velocity $\vec{v}$ do not change with time. The statement that the velocity does not change means that it is constant in magnitude and direction. This is just the law of inertia stated a little differently.

Fifth, the third law, the law of action and reaction claims a little too much: two charged particles moving in each other's electro-magnetic field exert equal and opposite forces on each other but these forces are not along their line of centers. In the context of the relative motion of charged particles, the line-of-centers assumption of the third law can be relaxed and (the rest of) Newton's laws still apply.

In the final analysis then, there are according to Newton only two principles which govern the motion of all material bodies. This abstraction and generalization is no less than monumental.

The study of Newtonian mechanics and its applications (and incidentally, his theory of gravitation) dominated physics for about 200 years (1700 to 1900). It is noted here that upon application, these two principles which are basically simple and easily comprehended can become exceedingly intricate mathematically owing to the fact that the dynamics law which is the central statement of the theory is a three-dimensional, second order differential equation. The physics is easy; the mathematics can get difficult.

The question might be asked, "Is Newtonian mechanics dead now that relativity and quantum mechanics have come along?" The answer is, "Emphatically, no!"

Newtonian mechanics explains all observable phenomena which take place at speeds not comparable to the speed of light and which involve objects not so small as to be comparable to the size of atoms. Consider for example the famous Navier-Stokes equations which govern aerodynamics: they were developed in the first half of the twentieth century and are still studied today. They are $\vec{F} = m\vec{a}$ applied to systems one of whose constituents is a compressible gas.

Where does Newtonian mechanics break down? It is interesting to track some of the history of physics from the late 1800's.

At that time the study of wave phenomena was well under way. In connection with sound waves, it was found that under certain excitations, Newton's dynamics equation determined that each differential section of a material medium oscillated back and forth in such a way that a (sound) wave was transmitted through the material. Further it was determined that sound waves could not travel from one point to another without a material medium to transmit them, i.e., they could not be transmitted across a vacuum. Other waves such as water waves were also known to need a medium.

In general, waves were found to be a way to get energy from one place to another through a medium without having to actually transport the medium. (Each differential piece of the medium just wiggled back and forth by a small amount causing the next differential piece to do the same, etc.). It was known further that the speed with which the waves travelled through the medium were a *function of the properties of the medium* and that they travelled at that appropriate speed *with respect to the medium transmitting them.* In another arena, optics to be specific, it was determined (mistakably) that light was unmistakably a wave.

The inevitable happened: some wise guy rocked the boat and asked, "How does sunlight get from the sun to the earth? There's no medium to carry the waves! What is wiggling back and forth?" "Hmmmm...," said the physicists, "that is a good question. You are indeed a very wise guy." They thought, and then said, "Ah, but the answer is simple: empty space is filled with a medium capable of transmitting light waves; we simply have never observed it." They compounded their folly: "Let's call this medium the ether", they said. The wise guy retorted, "What are the properties of this ether of yours?" After an awkward silence, they answered, "We'll have to check that out. We'll get back to you on that."

In order to answer this question they attempted to determine some things about the ether: in particular, they wanted to know if the ether was stationary (they didn't know quite what that meant) and just allowed objects to move through it freely or did it move along with such an object, or was it something in between, i.e. was it something like a viscous fluid which was partially dragged along? Whatever the answer was going to be, they dubbed this property the *'ether drag'* and they set up experiments to measure it.

Three separate experiments measuring the speed of light under certain predetermined conditions proved respectively that:

1. *the ether was absolutely stationary and allowed objects to move through it freely;*

2. *the ether was partially dragged along by the movement of a material body through it;*
3. *the ether was completely dragged along by the movement of any material through it.*

Uh oh, there's trouble! With these contradictions, the logic could proceed only as follows.

Assumption: a medium called the ether capable of transmitting light waves exists in empty space.

Experimental results:

1. there is no ether drag;
2. there is partial ether drag;
3. there is total ether drag.

Conclusion: the assumption leads to contradictions and is therefore, false;

hence, there is no such thing as the ether.

Well, now there was a problem. The speed of light was found to be $c = 186,282$ miles/second and was, according to the knowledge of the time, supposed to travel at that speed **with respect to the medium through which it was traveling,** i.e. the ether. But if there is no ether, light must travel at that rate **with respect to any inertial frame of reference** (i.e., with respect to every non-accelerating system.) Why? Well if there is no ether, then there is no way to distinguish between two inertial frames even if they are moving relative to each other. This implies that whatever is a physical fact in one must also be a physical fact in the other.

Now things get relatively crazy. As you may have guessed, here comes the man-on-the-train-and-the-man-on-the-platform paradox.

Imagine that your friend is on a train as it passes you at a constant velocity of *v, (100,* say*)* miles/hour while you are standing on the station platform. Imagine further that you are both equipped with the apparatus necessary to measure the speed of light. The plan is for each of you to measure the speed of a light beam as it passes.

First note that whether you are on the platform or on the train, you are in an inertial frame of reference. Suppose that when he passes you on the platform, a beam of light is traveling in the same direction as the train and is coming from behind the train about to overtake it. Your friend is running away from the approaching light and therefore it would be expected that the speed of the beam as measured by him be *100* miles/hour less than the speed as measured by you. But the statement, "There is no ether," implies that the speed of light is the same in his frame of reference as it is in yours. Paradoxically, your friend cannot 'run away' from the oncoming light beam: no matter how fast he goes, the light will still catch up to and pass him at its characteristic speed, *c*, which is the same in all inertial frames of reference.

This 'thought experiment' of the man-on-the-train and the man-on-the-platform can be carried much further. In fact, it can be carried to the point where all the results of special relativity can be determined. In particular, the relativistic formulas for time-dilation and length-compression are among the simplest which can be so determined. As an illustration, the time duration between two events *A* and *B* as seen from the train will be compared to the duration between the same two events as seen from the station platform. The result will be the expression for the relativistic time-dilation that applies to time comparisons between the two systems.

Surprisingly enough, the analysis needs only three things:

1. *the old grammar-school standby 'distance equals rate times time';*
2. *Pythagoras' theorem for right triangles* $(c^2 = a^2 + b^2)$;
3. *the experimentally determined fact that there is no ether.*

Suppose then that your friend on the train is holding a stick of length l oriented vertically. Let A be the event that a light beam is initiated at the bottom of the stick directed toward its top; let B be the event that the beam returns to the bottom of the stick after being reflected at the top by a mirror.

Analysis As Seen From the Train

Figure I. 1a shows the vertical stick. The beam travels to the top of the stick where it is reflected back to the bottom, traveling a total distance $2l$. The time duration τ for the round trip is determined by:

$$distance = rate \times time$$
$$2l = c\tau.$$

Analysis As Seen From the Platform

Figure I. 1b is a diagram of the same two events as seen from the platform. Note that the relative speed between you and your friend is v (the speed of the train). From your viewpoint on the platform the beam of light travels along the slanted paths between its initiation at A, and its return, at B as depicted in **figure I. 1 b**.

Clearly, when viewed from the platform system, the light must traverse a longer distance than it does when viewed from the train system. But, since there is no ether, the speed of light is the same in the two (inertial) systems. It follows that the duration between the events as seen from the platform is longer than the duration between the same two events as seen in the train.

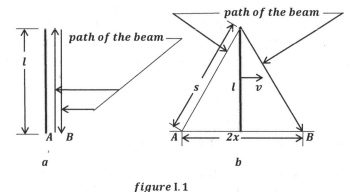

figure I. 1

Relativistic Time Dilation

It is a simple matter now to derive the time-dilation relation. Let s be the length of one of the slanted legs in **figure I. 1b** and let t be the time duration between events A and B as seen from the platform. Then t is determined by

$$distance \;=\; rate \;\times\; time$$
$$2s \;=\; ct$$

Also, the train has moved a distance $2x$ to the right in the time t so

$$2x = vt.$$

To collect all these facts, start with either one of the two triangles in *figure* I.1*b*. The following sequence of equations uses Pythagoras' theorem and the relations cited above.

$$s^2 = x^2 + l^2$$

$$\frac{c^2 t^2}{4} = \frac{v^2 t^2}{4} + \frac{c^2 \tau^2}{4}$$

$$\Rightarrow \quad \tau^2 = \left(1 - \frac{v^2}{c^2}\right) t^2$$

$$\Rightarrow \quad t = \frac{\tau}{\sqrt{1 - \frac{v^2}{c^2}}}.$$

The final result above is the relativistic time-dilation equation relating durations on the train to those on the platform. It says that time progresses at different rates in the two systems.

This theory (special relativity) was developed by Einstein in 1905. He made the mathematical requirement that the speed of light be the same in the two reference frames and that any discrepancies in length measurements or time duration measurements depend only on the relative velocity $\vec{v}$ and the speed of light c. All these requirements are a natural result of the fact that there is no absolute frame of reference (i.e., there is no ether). Put these ingredients into the mix, turn the mathematical crank and out pops the theory of special relativity. The above derivation illustrates this process.

The fact remains however, that Newton's laws had 200 years of experimental verification which could not be negated by the new relativistic mechanics. In this regard, it is important to consider the following: first, before the late nineteenth century, there were essentially no experiments carried out which involved speeds near the speed of light; second, at normal speeds, relativistic mechanics reduces to Newtonian mechanics. The bottom line is that relativistic mechanics explains (within experimental error) the 200 years of experiments which upheld the Newtonian theory because the speeds involved were so much less than the speed of light that relativistic mechanics and Newtonian mechanics were essentially the same. On the other hand, relativity explains a realm of phenomena involving larger speeds which Newtonian mechanics does not.

This must be the case if a new theory is to be considered more general than an older one: it must explain all the data explained by the older one in its realm of application as well as all the data in a new realm not explained by the older one. This is called the ***correspondence principle*** and it ensures that the newer theory is more general than the older. It can be concluded from the above remarks that relativistic mechanics can be accepted as the more general theory without disrupting 200 years of experimental evidence.

There are several results which were predicted by special relativity which are outside the realm of consideration by Newtonian mechanics. Salient among these is the equivalence of mass and energy, Einstein's famous equation

$$E = mc^2.$$

This equation germinated the investigation of atomic and nuclear energy. Since c is so large, it indicates that huge amounts of energy can be obtained from the transformation of a very small mass, **m**. [2]

There was another paper published by Einstein in 1905. This one was on the photoelectric effect and gave rise to the field of physics known as quantum mechanics. It proved (mistakably) that light was unmistakably a particle. "Glack!", you say. "How can one experiment prove that light is unmistakably a wave and another prove that it is unmistakably a particle?" (See previous discussion of waves.) We have yet another paradox on our hands. Its resolution brought about a complete change in the philosophies of physics on a metaphysical level.

The dilemma was that waves and particles are different things which have mutually contradictory properties. It is therefore impossible for an object to be both; if it is one, it cannot be the other. The only way out of the wave-particle paradox is to concede that light is neither, but has the **potential** to manifest itself either as a wave or a particle, depending on the conditions under which it interacts with another object. Subsequently it was found that this statement applied not only to light but also to every material object. It follows that no object can have intrinsic properties of its own; it can have only the potential for having properties and these manifest themselves only when there is an interaction of the object with something else. This implies that

> *physical objects are not observable; only*
> *interactions are observable.*

---

[2] Note that the amount of mass transformed into energy in the first A-bomb blast in the New Mexico desert was about the size of the head of a pin.

These concessions imply that when we attempt to observe an object, we actually observe its interaction with the device used to observe it. Looming on the horizon here is the Heisenberg uncertainty principle which states

> ***it is impossible to observe any physical object without altering its manifest properties in an unknown way.***

This principle which is derivable from our concessions is sometimes taken as the basis for quantum mechanics. It says essentially that it is impossible to play fly-on-the-wall looking at the universe as it parades by because the very process of looking at it changes it in some way.

Let's pin this down a little. Consider a bunch of billiard balls in movement on a billiard table, bouncing off each other and the sides of the table in some Newtonian fashion. In order to observe them, we turn on the light. Whatever the motion was before the light was turned on is now altered by the fact that the balls are being bombarded by photons. The effect is so miniscule that it is well within the experimental error of any measurement we might make.

But now suppose it is electrons we are observing instead of billiard balls The process of turning on the light to observe them has significantly altered their positions and velocities in an unknown way because the electrons are small enough so that the photon bombardment dramatically alters their previous motion. Thus the future positions and velocities are no longer deterministically predictable. The best that can be done is to find the probability that a given electron will be at a certain location with a certain velocity at some given future time. This is the condition of all of the elemental objects in the universe and hence, in some sense pervades its entirety.

Newtonian mechanics is called ***deterministic*** because it says that if the present state of every particle in the universe is known, then theoretically, its entire future (and past) can be calculated; quantum mechanics is called ***probabilistic*** because it says that it is impossible to observe the exact state of the universe and as a consequence only the probabilities of its future (and past) states can be calculated.

Einstein opposed quantum theory till his dying day even though he was in part its founder. In refutation of its wild assertions he said, "God does not play dice with the universe!" (He perhaps thought of quantum mechanics as his evil child who had a severe gambling problem.) .

After his 1905 papers, he went on to develop the general theory of relativity, (the relativistic theory of gravitation). This theory is the most satisfying field theory ever to be devised.

An account of the turbulent story of twentieth century physics and the wild conclusions of quantum theory is presented in the book, ***IN SEARCH OF SCHRÖEDINGER'S CAT***, John Gribbon, Bantam, 1984.

As implied by the examples above, the effects of quantum mechanics become significant when we deal with objects comparable to or smaller than the size of atoms. As in the case of relativistic mechanics, the equations of quantum mechanics reduce to those of Newtonian mechanics when larger objects are considered. Also, there were no experiments involving such small particles until the late 1800's, so quantum mechanics did not disrupt the 200 years of Newtonian success. In addition to the experiments of that time, it explains more recent experiments concerned with atomic and nuclear physics which cannot be explained by Newtonian mechanics. It therefore satisfies the correspondence principle and can be taken as a more general theory.

It would seem that the twentieth century left us in pretty good shape: Newtonian mechanics with its success as a complete world view, relativistic mechanics which legitimately extrapolates this world view to include situations where velocities comparable to the speed of light are involved and quantum mechanics which claims another legitimate extrapolation of the Newtonian world view to include situations involving objects comparable to the size of atoms and smaller. After these monumental achievements, it might be expected that physics could coast for a while exploiting these new finds until another major problem presented itself.

But physicists are intellectually restless and tend away from coasting. They conjectured, "There must be a single theory which reduces to Newtonian mechanics under 'normal' circumstances, to quantum mechanics when the objects of interest are atomic and/or smaller in size and to relativistic mechanics when velocities comparable to the speed of light are involved". Such a theory could address problems involving high-speed atomic particles.

They soon realized however that the new major problem was already contained in the advances mentioned above: not only did the search for this all-inclusive theory fail, but the unfortunate result of the attempts to find it was the conclusion that relativity and quantum mechanics were incompatible - they could not both be correct. Sad, but true. But...

...on the brighter side, many physicists currently working on what is called ***string theory*** believe that the issue described above will finally be resolved, not by extrapolating any of the existing theories but rather by inventing a bold new theory which could possibly accommodate all three. But that's another story.

When one considers that relativity and quantum mechanics have each demonstrated success as a theory in its own sphere of application and that each has already generated and still promises to generate exciting investigations and results, he can conclude only that Einstein left us a most beautiful puzzle.

From this point on, attention will be confined to Newtonian mechanics which is exciting in itself. It is an extensive world view which offers a vantage point to interpret and understand almost all physical phenomena which are normally observable.

It is the purpose here to demonstrate that all the laws of mechanics are the direct result of the basic dynamics equation

$$\vec{F} = m\vec{a}$$

and the law of action-reaction. It is assumed only that these laws apply to point masses (which is an idealization).

In **Chapter I**, such quantities as momentum, impulse, work, kinetic energy, etc. are defined and discussed in relation to the dynamics equation. In **Chapter II**, it is shown that the results for point masses can be generalized to arbitrary extended objects, solid, liquid or gas, in agreement with the starting point of Newton's theory. **Chapter III** presents some applications of the material already covered. Point masses are considered once again in **Chapter IV** in order to introduce the basics of circular motion. The relations between the parameters of circular motion are derived directly from the dynamics equation. This chapter also presents their generalization to arbitrary extended bodies. The subject of **Chapter V** is the application of the previous chapters to the analysis and solution of problems in mechanics. A description of Newton's theory of universal gravitation and an introduction to the theory of fields constitutes the subject matter of **Chapter VI**, which concludes this presentation.

Throughout the book, no assumptions are made except that the dynamics equation and the law of action-reaction apply to point masses. In the final chapter, the gravitational force field with its inverse square dependence is introduced. The form of this field does not constitute an additional assumption about the dynamics any more than does the introduction of the spring force $F = -kx$. These are simply forces whose forms were experimentally determined. They present new conditions under which objects still move according to the dictates of Newton's second and third laws.

CHAPTER 0

Preliminary Considerations

0.1 INTRODUCTORY REMARKS

Mechanics is the study of the motion of material objects under the influence of external forces. The description of such a motion clearly requires the mathematical ability to represent the position and orientation of the body in question during the course of its movement. This is a fairly complex and intricate problem in space geometry; its study is referred to as the science of **kinematics.** If the object in question is a rigid body for example, a minimum of six parameters is required; if the object is a fluid, it is necessary to specify the position and orientation at each point of the fluid's extent, that is, six parameters for each differential portion. In this latter case a complete description of the motion also necessitates tracking the shape and volumes changes of each differential piece.

The rigid body case has been studied extensively using the three coördinates of the center of mass to fix the position and the three Euler angles of rotation to fix the orientation. The gyroscope, a device sometimes used for navigation but more often as part of a feedback system used to stabilize the motion of ships and aircraft is a mechanical system which lends itself to analysis using these parameters. The equations of motion which result from this description allow for a mathematical determination of some of the subtle and unexpected behaviors which it can exhibit.

Be aware however that any number of parameter sets can be devised to describe any given mechanical system; and know further that the particular parameters chosen for the

1

task can render the problem dramatically more or less difficult to analyze.

Once the parameters are chosen, the **dynamics** of the system can be approached by writing the equation(s) of motion which relate the motion of the body to the forces which empower it. Equations which represent this relation are referred to as **dynamics equations.** They imply that the forces are causes whose effects are the ensuing motion. In this presentation, we will be concerned almost exclusively with Newton's famous dynamics equation

$$\vec{F} = m\,\vec{a}\,.$$

This equation, whose application to various situations germinates the bulk of this book, is a second order differential vector equation: 'vector', because the acceleration $\vec{a}$ and the force $\vec{F}$ appearing in the equation are vector quantities; 'second order', because the acceleration $\vec{a}$ is the second time-derivative of the position vector. (Keep in mind that the solution to a mechanics problem often consists of knowing the position and orientation of a mass as a function of time.)

It becomes apparent that some rudimentary knowledge of both vector algebra and differential and integral calculus would greatly enhance the understanding of the first lectures in mechanics. It is unfortunately the case however that in most instances, the course which presents this material is taught simultaneously with the first mechanics course. The result is that the material which could add so much understanding at the onset of the mechanics course lags by a month or two. In many circumstances this situation is tolerated and made to work but it is the opinion of this author that its effect on the student's understanding of the basic principles of mechanics is more detrimental than is appreciated.

The current section is inserted into the book to address this issue. It is the intent here to equip the student and/or the intellectually curious with a minimal but sound foundation for the first concepts of Newtonian mechanics.

To this end, vectors and the parallelogram rule for their addition are introduced in the next section. Then the representation of vectors in a Cartesian coödinate system is described followed by a list of a few properties of vectors. The topic of vectors is developed further in *section* 3.2.6, where there is a discussion of what is meant by a vector operation together with the definitions of the dot-product and the cross-product of two vectors. In *section* 3.2.7, there is a brief allusion to vector calculus.

In reference to calculus, there is a discussion of limits, derivatives, differentials and integrals: these descriptions are loose and devoid of rigor in the hope that the associated concepts be communicated without bogging them down in details. The discussion is directed at paving the way for an understanding of the basic implications of Newton's law of dynamics. This is followed by a geometric interpretation of the delta process and the expression for finding the first derivative of a given function. The section ends with a discussion of the fundamental theorem of integral calculus.

0.2 VECTORS

0.2.1   General Remarks

The playground for Newtonian mechanics is a three-dimensional Euclidean space. Many of the objects of interest like positions, velocities, accelerations, forces, etc., are not simple numbers: they are quantities which have directional properties in addition to their numerical values. Vectors, which have two properties, namely, magnitude (length) and direction were devised specifically to represent such quantities. For example, the velocity of an object traveling at 10 meters/second is represented by an arrow whose length represents the number ten and whose direction corresponds with the direction of motion. This simple, direct and seemingly innocent association can be developed further whereupon it gives rise to the elaborate algebra of vectors and matrices.

0.2.2   Addition and Subtraction of Vectors

The present purpose is to describe the rule for adding vectors and to indicate at the same time that the definition 'makes sense' when applied to a real situation. The notation $|\vec{A}|$ is used to represent the length of the vector $\vec{A}$ (i.e. if $\vec{A}$ represents a velocity vector then $|\vec{A}|$ represents the magnitude of the velocity, or the speed).

Figure 0.1 refers to the following situation. Suppose you and a friend are traveling together at a velocity $\vec{A}$. Then it follows that if you are at a point P at time $t = 0$, you will travel a distance of $|\vec{A}|$ units in the direction of $\vec{A}$ in one second placing you and your friend at the point Q at time $t = 1$.

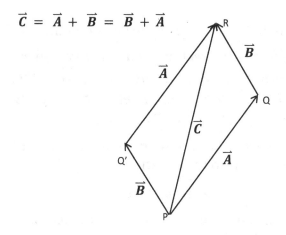

$$\vec{C} = \vec{A} + \vec{B} = \vec{B} + \vec{A}$$

figure 0.1

Parallelogram Rule for Vector Addition

Now consider a slightly different situation: suppose that you and your friend are together at point **P** and that you are traveling with velocity $\vec{A}$ as before while your friend is traveling with a velocity $\vec{B}$ with respect to you. Then after one second, he will be displaced from you by a distance $|\vec{B}|$ in the direction of $\vec{B}$. Reference to the figure indicates that while you end up at point **Q** as before, your friend ends up at point **R**. He has traveled from **P** to **R** in one second indicating that his net velocity is represented by the vector $\vec{C}$. Note that the situation was contrived so that conceptually, your friend's net velocity $\vec{C}$ can be considered as the (vector) sum of the two separate

5

velocities $\vec{A}$ and $\vec{B}$. Vector addition is defined in keeping with these ideas.

Reference to the figure reveals the vector addition rule:

> *to form the vector sum of two given vectors $\vec{A}$ and $\vec{B}$, place the tail of $\vec{B}$ at the head of $\vec{A}$. the vector sum, $\vec{C}$ then goes from the tail of $\vec{A}$ to the head of $\vec{B}$.*

Note that if one goes from P to R along the right path PQR, this definition leads to the equation

$$\vec{A} + \vec{B} = \vec{C};$$

whereas, if one goes from P to R along the left path $PQ'R$, then this definition leads to the equation

$$\vec{B} + \vec{A} = \vec{C}.$$

The definition for the addition of two vectors therefore leads immediately to the fact that addition is commutative, that is,

$$\vec{A} + \vec{B} = \vec{B} + \vec{A}.$$

It follows easily from this that given any finite number of vectors to be added together they can be reordered and grouped together (by the use of parentheses) in any way without altering the sum.

The above definition of vector addition is called the *parallelogram rule.*

The additive identity (i.e., the zero vector) in this algebra is the vector $\vec{0}$ which satisfies the equation,

$$\vec{A} + \vec{0} = \vec{A},$$

for *every* vector $\vec{A}$. If this requirement is considered in conjunction with the parallelogram rule for addition, a little thought reveals that this condition can be satisfied only if $\vec{0}$ is the vector whose magnitude (length) is zero. This vector has no directional properties.

The existence of the vector $\vec{0}$ equips us to find the vector $(-A)$ for any given vector $\vec{A}$: to this end, we seek a vector $\vec{X}$ such that

$$\vec{A} + \vec{X} = \vec{0}.$$

Then $\vec{X}$ can be taken as the negative of $\vec{A}$. The addition rule for vectors determines $\vec{X}$ to be the vector $\vec{A}$ with its head and tail reversed.

The difference of two vectors is then defined by the equation

$$\vec{A} - \vec{B} = \vec{A} + (-\vec{B}),$$

i.e., the vector difference $\overrightarrow{(A} - \vec{B})$ is found by adding the vector $(-\overrightarrow{B})$ to the vector $\vec{A}$.

0.2.3   Vectors and Cartesian Coördinate Systems

It is sometimes expedient, convenient and/or necessary to express a vector in a particular coördinate system. The simplest such system is one with three axes which point in three mutually perpendicular directions labeled the x-, y- and z-directions. This is called a *Cartesian coödinate system.* Any given vector $\vec{A}$ is expressed in this system by writing it as the sum of three separate vectors, one in each of the three major directions. In order to do this, three auxiliary *unit vectors* (vectors of length one) are defined: $\vec{\imath}, \vec{\jmath}$ and $\vec{k}$ which point in the x-, y- and z-directions respectively. The details of expressing $\vec{A}$ in this way follow.

7

Let A_x be the projection of $\vec{A}$ onto the x-axis. Then $A_x\vec{i}$ is a vector of length A_x in the x-direction. This is called the x-component of $\vec{A}$. Let A_y and A_z be the y- and z-projections respectively, and form the vector

$$\vec{A} = A_x\vec{i} + A_y\vec{j} + A_z\vec{k}.$$

This last expression exhibits the original vector $\vec{A}$ as the vector sum of its three component vectors.

As an exercise, draw a representation of a three dimensional coödinate system and show the given vector $\vec{A}$ together with the three vectors on the right side of the above equality. Convince yourself that according to the previous definition of vector addition, the sum of the three component vectors is equal to the given vector $\vec{A}$.

As another exercise, let

$$\vec{A} = A_x\vec{i} + A_y\vec{j} + A_z\vec{k}$$

and $$\vec{B} = B_x\vec{i} + B_y\vec{j} + B_z\vec{k}$$

be two given vectors and form their sum,

$$\vec{C} = \vec{A} + \vec{B}.$$

Once again, draw a diagram and convince yourself that

$$C_x = A_x + B_x$$
$$C_y = A_y + B_y$$

and $$C_z = A_z + B_z.$$

This indicates that vectors can be added componentwise, i.e.

$$\vec{A} + \vec{B} = (A_x + B_x)\vec{i} + (A_y + B_y)\vec{j} + (A_z + B_z)\vec{k}$$

The following is a list of basic facts about vectors that follow immediately:

1. *a vector is zero if and only if each of its components is zero;*
2. *two vectors are equal if and only if each of their corresponding components are equal;*
3. *a vector is constant if and only if its magnitude is constant and it does not change direction;*
4. *a constant vector is not involved in any differentiation process; it is carried along the same as a scaler;*
5. *any derivative of a vector of constant magnitude is normal to the vector.*

The first three assertions above are obvious. The fourth and fifth may need some explanation.

Pertaining to the fourth, the mathematics of differentiation is the mathematics of changes. Any quantity that does not change cannot be involved in a differentiation process: it assumes the role of a constant.

Pertaining to the fifth, any (component of) change in a vector in its own direction indicates a change in its length. If a vector has constant magnitude, then any change must be normal to its own direction.

0.3 BASIC CONCEPTS OF CALCULUS

0.3.1 Introductory Remarks

In the mathematics of pre-calculus, the operations at our disposal are addition, subtraction, multiplication, division, powers and roots and combinations of these. For example, the expression

$$\sqrt[3]{x^2 + 1} \, ,$$

represents the quantity obtained when the number **1** is added to the square of x and the cube root of the result is taken. In the process of evaluating the expression, we have raised to a second power, added and taken a cube root. All of these operations are familiar from our earlier bouts with mathematics.

Calculus adds one more operation: that of taking **limits.** It is not the purpose here to develop the theory of limits with any mathematical rigor. This aspect of the topic is adequately covered in any course in elementary calculus. The current purpose is rather to establish their importance as a basis for the calculus operations of differentiation and integration. Without limits these operations which pervade all analyses cannot be accepted as valid.

What is presented below is a description of the nature of limits and some of the motivations that require their study. This is accompanied by a brief discussion of the underlying theory which establishes them as a new operation. It will be found that allowing limits into our list of operations gives rise to some new types of quantities with their own notations. These new quantities include differentials, derivatives, anti-derivatives and integrals. It is a further purpose of this section to present the broadest descriptions of these associated concepts and to introduce the notations that accompany them.

0.3.2 Limits

The simplest analyses in calculus often require us to deal with indeterminate forms. For example, to find the derivative of a function $f(x)$ (this is a basic calculus operation which will be defined and discussed shortly), one deals with expressions like $\frac{\Delta y}{\Delta x}$ where the quantity Δy depends on the quantity Δx, and whose dependency is such that $\Delta y = 0$ when $\Delta x = 0$. The problem arises when the analysis requires us to associate some quantity with the ratio $\frac{\Delta y}{\Delta x}$ when $\Delta x = 0$.

This produces a dilemma because when $\Delta x = 0$,

$$\frac{\Delta y}{\Delta x} = \frac{0}{0},$$

which is an indeterminate form. The entire process of finding the derivative of a given function is crippled at this point unless this situation can be remedied.

Since the ratio cannot be defined at the point of interest we do the next best thing: we investigate the behavior of the ratio in the *vicinity* of $\Delta x = 0$. This is done by letting $|\Delta x|$ be very small and letting it sneak up on 0 by going through some sequence of values, each closer to 0 than the previous. This sneaking-up process is indicated by the notation

$$\Delta x \to 0$$

which is read, "delta x goes to zero" or, depending on the syntax where the expression occurs, "as delta x goes to zero". It must be stipulated however, that Δx never actually equal 0 because that would simply throw us back into the dilemma which started all this.

The question then arises, "How close to zero?" But that's like asking, "What is the smallest positive number?" There is no definitive answer to this last question because no matter what we pick as the smallest positive number, it can be (for example) divided by **2** producing an even smaller one. In fact, if we were to type a decimal point followed by a sequence of **0′s** long enough to reach around the earth and put a **1** at the end, know that $|\Delta x|$ can be made smaller than the resulting number. The attempts to verbalize this extreme, use expressions such as 'arbitrarily close to zero', 'infinitesmal in size' and the like. The idea is that the ratio in question seems to have some definite value when Δx is near zero, i.e. as

$$\Delta x \rightarrow 0, \qquad \text{the ratio} \qquad \frac{\Delta y}{\Delta x} \rightarrow L,$$

where L is a quantity which can be found. If such an L exists, it is required that it be unique; otherwise the results become ambiguous.

Recall that this problem arose in connection with an attempt to find the derivative of $f(x)$ with respect to x (yet to be defined). For the time being, without understanding what a derivative really is, we introduce the notation used to indicate the process just described:

$$\frac{df(x)}{dx} = \lim_{\Delta x \to 0} \frac{\Delta f}{\Delta x} = L(x),$$

where the expression on the left is one of the common notations for the derivative of $f(x)$ with respect to its argument x. The notation

$$\lim_{\Delta x \to 0} Z$$

is used to express what was described above: it means and is read, "the limit as $\Delta x \rightarrow 0$ of the quantity Z". In connection with the x which occurs on the right, note that

x is not involved in the limiting process. The fact that L may still be a function of x after the differentiation process has been completed indicates that the derivative may very well have different values at different values of x.

The previous paragraph applies only to those cases where L exists and is unique. If these requirements are not met, then the derivative of the function f cannot be defined.

The foregoing discussion illustrates that the validation for the entire subject of differential calculus requires that the theory of limits be put on a solid logical foundation.

Finding derivatives is not the only context which requires the evaluation of limits. Suppose for example, it is required to find the area A bounded above by some curve $f(x)$, below by the x-axis, on the left by the line $x = a$, and on the right by the line $x = b$. This area is called the integral of $f(x)$ and is notated

$$A = \int_a^b f(x)dx.$$

The problem is solved as follows.

The x-axis between $x = a$, and $x = b$ is divided into a large number n of pieces. (See *figure* 0.2.) Then each piece is taken as the width of a tall, thin rectangle. The heights of the rectangles are determined so as to just touch the lowest point of the curve in the current region. The collective area of these rectangles underestimates the area A. Designate this underestimate, A_u.

Using each piece of the x-axis as the width of a tall, thin rectangle once again construct a second set of rectangles whose heights are determined so as to touch the highest point of the curve in the current region. The collective area

13

of this second set of rectangles overestimates the area A. Designate this overestimate, A_o

Then we can write

$$A_u \leq A \leq A_o.$$

Now divide each rectangular width in half thus doubling the number n of rectangles. If the heights of the new rectangles are determined in the same way as those of the previous prescription, then each of the quantities A_u and A_o becomes a better estimate of A. (Exercise: make a diagram indicating how these 'better estimates' come about.)

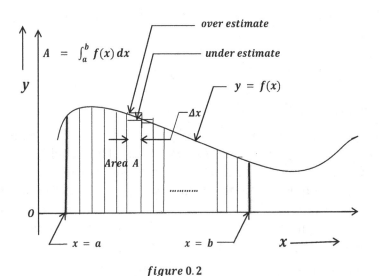

figure 0.2

Area under a Curve

We repeat this process indefinitely, doubling the number of divisions along the x-axis at each stage, and letting n

increase without limit. Let L_u and L_o if they exist be defined by

$$L_u \quad = \quad \lim_{n \to \infty} A_u$$

and

$$L_o \quad = \quad \lim_{n \to \infty} A_o \,.$$

Suppose now that L_u and L_o are equal, i.e., both A_u and A_o have the same limit L. All during the limiting process, the true area A was sandwiched between the underestimate and the overestimate. Then, since these two estimates approach the same limit L, it follows that

$$\lim_{n \to \infty} A_u \quad \leq \quad A \quad \leq \quad \lim_{n \to \infty} A_o$$

$$L_u \quad \leq \quad A \quad \leq \quad L_o$$

$$L \quad \leq \quad A \quad \leq \quad L$$

$$\Rightarrow \quad A \quad = \quad L$$

(The symbol $\Rightarrow$ means "implies". It indicates that what comes next follows from what precedes.) Hence as in the case of differential calculus, the validation of integral calculus also requires that the theory of limits be put on a solid logical basis.

Hence limits are the foundation on which calculus depends. This fact prompts us to carry this investigation a little further.

Focus attention once again on the expression

$$\lim_{x \to a} f(x) \quad = \quad L$$

Recall that the question was asked before, "How close do things have to be?" referring to the distances from x to a, and from $f(x)$ to L and the magic answer was 'arbitrarily close' or 'infinitesimally close'. We can make x arbitrarily close to **a**, i.e. we can pick a positive number **δ,** infinitesimal in size if need be, so that the values of **x** under consideration are in the range (**a** − **δ**) to (**a** + **δ**). We will call this a **δ**-range around **x** = **a**. Similarly, we call the range (**L** − **ε**) to (**L** + **ε**) an **ε**-range around **L**. It is easy to show that given a positive number **ε,** no matter how small, **IF** it is always possible to find **δ** such that for all **x** in the **δ**-range around **a**, **f**(**x**) is in an **ε**-range around **L**, i.e. if

$$|a - x| < \delta \quad \Longrightarrow \quad |L - f(x)| < \varepsilon$$

then the limit **L** in the above equation exists and is unique. (The last condition is read "if **x** is closer to **a** than **δ** implies that **f**(**x**) is closer to **L** than **ε**...". This is logically equivalent to "if the condition that **x** is in a **δ**-range of **a** implies that **f**(**x**) is in an **ε**-range of **L**...").

Proof:

L is certainly a candidate to be a limit since it is in every **ε**-range around itself.

To show that **L** is unique, suppose there is another limit **L'** different from **L**. Let **ε'** be the difference

$$|L - L'| = \varepsilon'.$$

Pick some positive **ε** smaller than **ε'** and find a **δ**-range around **x** = **a**, whose resident **x**'**s** are such that **f**(**x**) is in an **ε**-range around **L**. (It is the condition of the problem that such a **δ**-range can always be found.) Since **ε** was chosen to be less than **ε'** this **ε**-range does not include **L'**. Therefore **L'** cannot be a limit. Thus the condition cited above renders **L** the **unique** limit.

This completes the proof.

This section is concluded with a few remarks.

While the theory of limits empowers just about all of calculus, it is seldom necessary to deal with limit problems or to justify an operation by showing that any associated limit exists. The truth of the matter is that most all functions of interest in the application of calculus to practical situations are continuous, bounded and all those good things and as such do not present limit problems. But the theory of limits must be there anyway just as your parents must be there to pay for your education even though they do not actually go to class with you. The theory of limits is there to ensure the validity of our results even though we may take a thousand derivatives and evaluate a thousand integrals without ever giving a thought to the fact that implied in every one of these operations was the use of limits and that each result obtained was validated by them. Under normal circumstances limit problems just do not present themselves as an issue.

In the previous discussions, it might be noticed that the complexity of the language syntax necessary to speak about limits often exceeds the complexity of the ideas expressed. Be aware that the ideas are simple; it is the process of expressing them in language that makes them seem complex. This is due at least in part to the fact that we are talking about such things as sets of points in certain vicinities. The language of topology is better suited to present such ideas and in fact, when spoken about from a topological point of view, the topic of limits (also the theory of continuity) can be presented in a much more natural way.

0.3.3 The Derivative of $f(x)$

In order to describe what a derivative is, it is helpful to understand the rather universal use of the Greek letter Δ (delta). It is used to represent a **change** in the quantity which it modifies. More explicitly, Δx does **not** represent some quantity Δ times another quantity x: Δx is a single independent variable representing a change in x.

A derivative is a basically simple thing: it is the change of $f(x)$ per unit change in x. Imagine that you move along the x-axis from some point x to a nearby point $(x + \Delta x)$ and you notice that as you do so $f(x)$ changes by an amount $\Delta f(x)$, then the derivative of $f(x)$ with respect to x is the result of comparing these changes by taking their ratio $\Delta f(x)/\Delta x$. Thus the derivative answers the question. "as x changes by a small amount, how many times faster does $f(x)$ change?". Once again, it is the change in the function $f(x)$ per unit change in its variable x.

This description defines an algebraic process for finding the derivative of a given function $f(x)$. It is called the "delta process" and is described as follows. (Reference to *figure* 0.3 offers a geometric interpretation.)

$$\frac{df(x)}{dx} = \lim_{\Delta x \to 0} \frac{\Delta f(x)}{\Delta x}$$

$$= \lim_{\Delta x \to 0} \frac{f(x+\Delta x)-f(x)}{\Delta x}$$

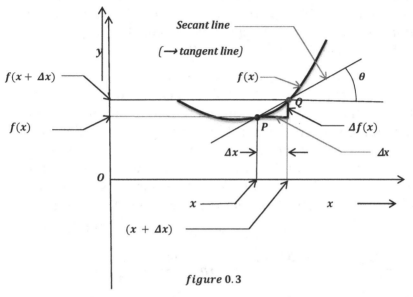

figure 0.3

The Derivative of a Function

The diagram in the figure shows the graph of some given function $f(x)$. We pick an arbitrary but unspecified point x along the x-axis and propose to find the derivative of $f(x)$ at this value of x.

Note that the change in the function as you moved from x to $(x + \Delta x)$ is

$$\Delta f(x) = f(x + \Delta x) - f(x) .$$

This change is compared to the change in x by considering the ratio

$$\frac{\Delta f(x)}{\Delta x} = \frac{f(x + \Delta x) - f(x)}{\Delta x}.$$

This is almost the result we are after. The only problem is that the ratio of the change in $f(x)$ to the change in x is not constant over the Δx region. We want this ratio **at the point x.** There is one thing left to do, namely, take the limit of the above expression as $\Delta x \rightarrow 0$.

Reference to the **figure 0.3** shows that in this limiting process, the point Q gets closer and closer to P until in the limit they are separated only infinitesimally. Simultaneously, the secant line which intersects the curve at both P and Q becomes the line tangent to the curve at P. After the limiting process, we have what is called the derivative of the function $f(x)$. Note that this is the rise over the run or the **slope** of the tangent line. The derivative denoted $\frac{df(x)}{dx}$ or simply $\frac{df}{dx}$ is therefore given by the formula

$$\frac{df(x)}{dx} = \lim_{\Delta x \to 0} \frac{\Delta f(x)}{\Delta x} = \lim_{\Delta x \to 0} \frac{f(x + \Delta x) - f(x)}{\Delta x}.$$

Note that during the above discussion, it was shown that the derivative of $f(x)$ is the slope of (the line tangent to) $f(x)$ at the point x. This can be taken as an alternate definition of the derivative. Also note that the slope is a function of x, i.e. in general, the slope of $f(x)$ changes as x changes.

As an example of finding a derivative using the above prescription, consider the function

$$f(x) = x^3$$

Then, using the formula presented above,

$$\frac{df}{dx} = \lim_{\Delta x \to 0} \frac{f(x + \Delta x) - f(x)}{\Delta x}$$

$$\frac{df}{dx} = \lim_{\Delta x \to 0} \frac{(x + \Delta x)^3 - x^3}{\Delta x}$$

$$\frac{df}{dx} = \lim_{\Delta x \to 0} \frac{3x^2\Delta x + 3x(\Delta x)^2 + (\Delta x)^3}{\Delta x}$$

$$\frac{df}{dx} = \lim_{\Delta x \to 0} (3x^2 + 3x\Delta x + \Delta x^2)$$

$$\frac{df}{dx} = 3x^2$$

This is the result: if $f(x) = x^3$, then the derivative of (or the slope of the line tangent to) $f(x) = x^3$ at some given value of x is $3x^2$.

The point of the previous discussion is to illustrate the application of the delta process as a formula for finding the derivative of a given function.

The list of applications of derivatives to analyses of various scientific endeavors is endless. They have pervaded almost every such effort since the time of their discovery in the time of Newton and they continue to do so at the present time.

0.3.4 Differentials

Some of the expressions which occur when taking a derivative need some discussion, namely,

$$\frac{\Delta f}{\Delta x} \quad \text{and} \quad \frac{df}{dx}.$$

The first of these is simply an algebraic ratio of two small (but not infinitesimal) numbers. Once the limit of this expression is taken, the result is the second expression which is the derivative. The second expression *conceptually*, is a ratio of two infinitesimal numbers.

21

However, in the absence of an arithmetic that applies to infinitesmal numbers, it is not strictly legitimate to treat the 'numerator' and 'denominator' of the expression for a derivative as separate quantities: the whole symbol represents one quantity, the derivative.

The unfortunate aspect of this situation is that when treated as separate small algebraic quantities, the 'numerator' and 'denominator' offer conceptual insights to those who are just learning calculus and attempting to use it for the first time to describe physical situations. For example, ds might be identified as a differential distance along a path of motion, and $dt,$ as the differential time it takes to traverse the distance ds: then if v is the speed, it is natural to write

$$ds \;=\; v\,dt\,,$$

which is just 'distance equals rate times time', an old grammar school standby. Also, we might deduce from this, ('dividing' by dt) that

$$\frac{ds}{dt} \;=\; v$$

which exhibits the speed as the time-derivative of the distance. This last relation is valid: the (instantaneous) velocity *is* the change in position per unit change in time.

Thus the 'numerator' and 'denominator' of a derivative expression carry helpful connotations and for this reason will be treated here as if they are algebraic in nature. This will work as long as the derivative is defined at the point(s) in question and for our purposes, this is always the case.[3] At those points where the derivative exists, the

---

[3] The practice of treating the 'numerator' and 'denominator' of a derivative as if they were algebraic quantities can be legitimized as follows. Recall that it was determined that the derivative of a function $f(x)$ evaluated at the point $(x, f(x))$ is the slope of the line tangent to the curve at x. Choose a point

two equations above say the same thing: the first is the *differential* form; the second is the *derivative* form.

(Note however, that if you are in school, you may treat the 'numerator' and 'denominator' as algebraic quantities in your physics class but not in your calculus class.)

0.3.5 The Definite Integral

In section 0.3.2 on limits, the problem of finding the area under a curve was described briefly. The accompanying *figure* 0.2 is a representation of this description. The following is a more detailed look at this problem and is included here to impart some appreciation for the notation used to represent a definite integral.

Recall that the region of interest was broken up into a large number of rectangles to underestimate and overestimate the area in question. Then the number of rectangles was doubled by halving their widths, and better over- and under-estimates were obtained. The doubling process was repeated indefinitely. The limits were taken as the number n of rectangles approached infinity and it was seen that if the limits were sufficiently well behaved, the area in question could be determined.

---

on the tangent line near x and let $\Delta f(x)$ and Δx be the vertical and horizontal changes that occur *along the tangent line* as you move from x to the chosen point $(x + \Delta x)$. Then the ratio of these two changes is the derivative evaluated at x regardless of the distance Δx. (The ratio is the [constant] slope of the line tangent to the curve at x.) The chosen point can then be arbitrarily close to the point of tangency making it possible to bring all associated quantities arbitrarily close to their values when the real derivative is used. It follows that the 'numerator' and 'denominator' of a derivative can be treated as algebraic as long as the derivative exists at the point in question, i.e. as long as there is a unique line tangent to $f(x)$ at x.

This process will now be represented in mathematical symbols: when n is finite, the expressions are simply algebraic; when the limit as $n \to \infty$ is taken, the calculus notation replaces the algebraic. This new notation will be looked at in some detail as it has important connotations when associated with the symbols from which it came.

Divide the x-axis from $x = a$ to $x = b$ into a large number, n of segments. For simplicity, make all these segments equal. (This is not necessary but it's legitimate and it makes things easier.) Let Δx be the width of each of the rectangles. Then (the increments of x are)

$$\Delta x \;=\; \frac{b - a}{n}$$

There are now n box-widths laid out along the x-axis. Number the corresponding boxes $1, 2, ... n$, and focus attention on a typical one, say, the kth box. For the underestimate of the area, choose as the height of the kth box the smallest height above the x-axis which touches the curve. Call this height y_k. Then the (approximate) area A_k of this box is

$$\Delta A_k \;=\; y_k \Delta x$$

Adding the areas of these rectangles together, the underestimate of the area A is

$$A_u \;=\; (y_1 + y_2 + \; ... \; + y_n) \, \Delta x$$

$$A_u \;=\; \sum_{k=1}^{n} \Delta A_k \;=\; \sum_{k=1}^{n} y_k \, \Delta x$$

with a similar expression for the overestimate of A. (This last expression is algebraic in nature.) Assume that this underestimate approaches the real area A. We now take the limit and replace the above algebra notation with the calculus notation:

$$A = \lim_{n \to \infty} \sum_{k=1}^{n} \Delta A_k = \lim_{n \to \infty} \sum_{k=1}^{n} y_k \Delta x$$

As $n \to \infty$, the following three things happen:

$$\Delta x \quad \to \quad dx \quad \quad (differential\ of\ x)$$

$$y_k \quad \to \quad y(x)$$

$$\sum_{k=1}^{n} \quad \to \quad \int_a^b$$

The resulting calculus notation for the area A is

$$A = \int_a^b y(x)\ dx.$$

The last expression on the right is called the **definite integral** *of y(x)*; *'a'*and *'b'*are called the lower and upper limits of the integral, respectively; $y(x)$ is called the integrand; dx is a differential (infinitesimal) change in x.

Each part of the definite integral descended during the limiting process from an analogous part of the precedent algebraic expression. The differential dx descends from the finite $\Delta x's$ and can be thought of as the width of an infinitely thin box whose height is $y(x)$. Since we are conceptualizing boxes with infinitesimal widths, there is no longer a larger and smaller box constructed at the point x: the single value $y(x)$ descends from both the larger and smaller height (as long as $y(x)$ is continuous at that point). At the point x then, there is a differential contribution dA to the area A from a 'box' of height $y(x)$ and width dx:

$$dA = y(x)dx$$

The integral sign itself descends from the algebraic summation sign and represents the addition of all these 'boxes' to produce the whole area A. The summation over k (from $k = 1$ to $k = n$), is replaced by the stipulation that the areas of these differential 'boxes' be added together from $x = a$ to $x = b$.

Conceptually then, the definite integral can be thought of as the sum of an infinite number of infinitely thin boxes which add up to the area A.

$$A \;=\; \int_a^b dA \;=\; \int_a^b y(x)\,dx$$

Note that the area A has a specific numerical value. The word 'definite' in the designation ***definite integral*** refers to the fact that the integral has limits on it (namely, a and b) and represents a definite numerical value for A.

There is no rigor in the preceding conceptualization of the definite integral. It is simply a way of putting integration on more familiar ground. It offers some helpful insights especially when formulating a mathematical model to describe a physical situation.

0.3.6 The Indefinite Integral

In this section the relationship between integrals and derivatives will be examined. It will be found that integrating and differentiating are inverse operations: i.e. the derivative of the indefinite integral of $y(x)$ is $y(x)$. This fact is not at all obvious from the way in which integrals and derivatives were defined previously.

We construct a situation similar to the area problem of the previous section, but this time the upper limit of the integral will be a variable t. This is the definition of the indefinite integral:

$$A(t) \;=\; \int_a^t y(x)\,dx$$

The variable x now starts at $x = a$, as before and goes to some arbitrary but unspecified point $x = t$.. The t in parentheses on the left side of the above equation, is an explicit indication that the area A is a function of t (If a different t further to the right is chosen then $A(t)$ gets larger.)

Admittedly this type of dependence on t is slightly offbeat but nevertheless, it is legitimate. The integral in question has no specific numerical value and is called the ***indefinite integral of y(x).*** It is the intention here to find the derivative of $A(t)$ by a straightforward application of the delta process. It will be seen that the derivative of the integral under consideration is $y(t)$. In other words, if $y(t)$ is integrated and its integral is then differentiated, the result is $y(t)$. This will establish the fact that differentiation is the inverse of indefinite integration and visa versa.

Refer to ***figure 0.4*** in connection with the following steps which describe the application of the delta process to find the derivative $dA(t)/dt$.

$$\frac{dA(t)}{dt} = \lim_{\Delta t \to 0} \frac{A(t + \Delta t) - A(t)}{\Delta t}$$

$$\frac{dA(t)}{dt} = \lim_{\Delta t \to 0} \frac{\int_a^{t+\Delta t} y(x)dx - \int_a^t y(x)\, dx}{\Delta t}$$

Note that the numerator in this last expression is the area which includes the box on the right, minus the original area (up to $x = t$) which does not include the box on the right. The difference is just the area of the box on the right, i.e., the area from t to $(t + \Delta t)$:

$$\frac{dA(t)}{dt} \; = \; \lim_{\Delta t \to 0} \frac{\int_t^{t+\Delta t} y(x)dx}{\Delta t}$$

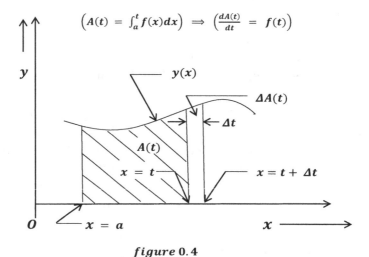

$$\left(A(t) = \int_a^t f(x)dx\right) \; \Rightarrow \; \left(\frac{dA(t)}{dt} = f(t)\right)$$

figure 0.4

Indefinite Integration and the Fundamental Theorem

As long as the curve is continuous at $x = t$, (i.e. it does not jump from one value to another at this point), we can write

$$\frac{dA(t)}{dt} \; = \; \lim_{\Delta t \to 0} \frac{y(t)\Delta t}{\Delta t}$$

$$\frac{dA(t)}{dt} \; = \; y(t)$$

What has just been shown is that the rate of change of the area under the curve with respect to a change in the upper limit is the curve itself. The last equation exhibits $A(t)$ as the anti-derivative of $y(t)$. Hence we have...

> *The Fundamental Theorem of Integral Calculus: if A(t) is the indefinite integral of y(t), then y(t) is the derivative of A(t).*

In other words, ***indefinite integration is identical to anti-differentiation.*** There is, however one respect in which some care must be taken. It arises because the derivative of a constant is zero. (This can of course be shown by applying the delta process to $f(x) = c$ where c is a constant but this is not necessary because the proof is contained in the English language: a constant is constant. Hence its rate of change with respect to any variable is zero. It simply doesn't change.) It follows then that if $F(x)$ is a function whose derivative is $f(x)$, then the derivative of the more general function $F(x) + C$ where C is any constant is also $f(x)$. Therefore, if $f(x)$ is to be integrated, the possibility of having lost a constant must be taken into account. Hence the most general value of the indefinite integral of $f(x)$ is

$$\int f(x)\, dx \quad = \quad F(x) + C,$$

where C is an arbitrary constant. C is called the ***constant of integration.***

More simply, note that adding the arbitrary constant C merely translates $F(x)$ up or down vertically without changing its shape. This operation does not affect the derivative (which is the slope of the curve).

Incidentally, the indefinite integral is usually written as in this last equation, i.e. without any limits indicated.

0.3.7 Evaluating a Definite Integral

There is a small matter to be cleaned up, the evaluation of a definite integral. All the work has been done already. It needs only to be collected. Consider the integral

$$\int_a^b f(x)dx$$

Let r be some value of x to the left of $x = a$. Then (note the limits on the integrals)

$$\int_a^b f(x)dx = \int_r^b f(x)\,dx - \int_r^a f(x)dx$$

This last equation comes from adding and subtracting areas under the curve $f(x)$. As before the area is a function of the upper limit. It follows from the above that

$$\int_a^b f(x)dx = A(b) - A(a)$$

where the function A is the indefinite integral of f.

This last equation is the result: to evaluate a definite integral, find the indefinite integral of the given function, evaluate the result at the upper and lower limits and subtract the latter from the former.

The following is noted with regard to the processes of differentiation and integration. The delta process described earlier offers a deductive method for finding the derivative of an arbitrary given function. It can usually be applied without much difficulty. A deductive method for integrating however almost invariably involves the evaluation of sums much too cumbersome to deal with. As a result, there are extensive tables listing the integrals of many algebraic, trigonometric, exponential and logarithmic forms. Additionally, any course in elementary calculus includes *'methods of integration'* as part of its subject matter. This section of such a course consists of numerous tricks by which almost any integrand can be reduced to one of the tabulated forms.

As a final remark, we state that what appears under the integral sign is always the differential of something,

$$\int_a^b f(x)\, dx \;=\; \int_a^b d[Q(x)] \;=\; Q(b) - Q(a),$$

whether or not the 'something' ($Q(x)$ in this case) is known. The expression to the right of the first equal sign in the above equation exhibits the integral sign $\int_a^b$ and the differential indicator, $'d'$, as inverse operations which cancel one another out. In other words, if you break Q up into an infinite number of differential pieces ($dQ's$) and then add them all back together, i.e. integrate as in the second term above, you get Q back again. It's a Humpty Dumpty story with a happy ending.

Take this one step further: if the differential dx is treated as an algebraic quantity, (see **section 0.3.4**) a practice which can be legitimized as long as $\frac{dQ(x)}{dx}$ is defined everywhere in the region of integration, (i.e., $a < x < b$) then the last equation can be written

$$\int_a^b f(x)\, dx \;=\; \int_a^b \frac{dQ(x)}{dx}\, dx \;=\; Q(b) - Q(a).$$

This is an identity which holds for any arbitrary values of a and b. It follows that the integrands in the two integrals must be equal. Hence

$$f(x) \;=\; \frac{dQ(x)}{dx}$$

thus exhibiting $f(x)$ as the derivative of its integral. This is just the fundamental theorem arrived at by a different method.

CHAPTER I

The Basic Implications of Newton's Laws

1.1 ORIENTATON AND IDEALIZATIONS

1.1.1 The Goal of Theoretical Mechanics

Theoretical mechanics purports to provide a detailed description of the positions and orientations of all physical bodies, how these quantities change with time and how the motions so described are related to the forces in play. In the case of quantum mechanics, this statement must be somewhat loosened, replacing "...the positions and orientations..." by "...the probabilities for every possible position and orientation...".The problem of describing the position and orientation of a body as a function of time during the course of its motion (apart from the forces acting on it) is a considerably difficult problem in itself: it is essentially a problem in space geometry and is referred to as the science of *kinematics*. The way in which this motion is related to the forces in play is called *dynamics.* Hence, $\vec{F} = m\vec{a}$ is a dynamics equation since it relates the force to the motion through the acceleration.

It is generally true that given all the forces acting on a body, the complete details of its motion can be determined by the application of Newton's laws. The converse is also true: given the complete details of the motion of the body, the forces on it can be determined. In the real world however, being given one or the other of these is a luxury which is not always available. For example, in an electrically charged gas (a plasma), the forces which act on one atom (or molecule) depend on the positions and velocities of every other atom (or molecule). Therefore the forces are not known until the motion is determined and the motion cannot be determined until the forces are known. It is generally true that the application of the simple concepts of Newton's dynamic principles leads quickly to horrendous mathematical intricacies.

To circumvent some of these intricacies which occur in many contexts, certain idealizations are introduced. One enters the world of massless and frictionless pulleys, frictionless surfaces, massless springs and strings, etc. The purpose of these idealizations is to focus attention on the current topic of study. They sometimes eliminate difficult problems which do not affect the results to any great degree. They also avoid long equations where the algebraic manipulations tend to obscure the physics. The use of idealizations is a valuable tool in the process of progressing from simple to difficult analyses and from rough to more refined results in a given analysis.

1.1.2 The Nature and Purpose of Idealizations

As mentioned above, numerous idealizations are introduced in a course of this nature. Know that in problems involving frictionless surfaces, your results are essentially correct if the surfaces are at least slippery; and in problems involving massless pulleys, your results are essentially correct as long as the mass of the pulley is small compared the other masses in the problem. Since a number of these idealizations are directed at neglecting the friction forces, it is helpful to know that in real-world problems involving small friction forces, the problem is often solved as if everything were frictionless to get a preliminary result which is then refined by taking the friction into account. In this context, the friction acts like an agent which causes a **perturbation** to the result already obtained.

Contrary to the above, the first idealization which we are going to make is not at all reasonable: we idealize a body as a mass all concentrated at a **single point.** This simplification is not even an approximation to reality. Nevertheless, it is a highly valuable assumption for a number of reasons. First, it allows for the Newtonian equations to be analyzed in their simplest form; second, it avoids the issues of the extent and shape of the body; third, it eliminates any complications introduced by rotational motion. This idealization would be next to useless if it weren't for the fact that it can be shown (and it will be in Chapter II) that

> **the assumption that Newton's laws govern the motions of point masses implies that they also govern the motion of arbitrary real extended bodies.**

The implication is that the difficulties avoided by this idealization can be incorporated into the developing theory a little at a time. Hence the approach to mechanics taken here is based on the one simple...

> **Assumption:** *Newton's laws of motion govern the motion of point masses.*

The remainder of this presentation is a skeletal view of what this single assumption implies.

1.2   THE BASIC RELATIONS IMPLIED BY NEWTON'S LAWS

1.2.1   General Remarks

In this chapter attention is focused on the reaction of a point mass to the application of an external force. The force will be viewed in this context as a **causal** element whose **effect** is the ensuing motion. Conditions and stipulations which tend to particularize this given situation will be avoided as much as possible to determine what can be said in general about the implications of Newton's dynamics law

$$\vec{F} \;=\; m\vec{a}.$$

It will be seen that the attempt to determine the motion of the mass motivates the definitions of four quantities that are important parameters in the analysis of any mechanics problem. They are:

1   the **impulse** delivered by the force;
2   the **momentum** of the mass;
3   the **work** done by the force;
4   the **kinetic energy** of the mass.

Additionally the partition of forces into two types called **conservative** and **non-conservative** will motivate the definition of another quantity:

5   the **potential energy**.

These quantities will be defined and described as we proceed with the solution of the dynamics equation. Understanding the nature of these five quantities and the relations between them is not only the content of this chapter but also the basis of all that follows.[4]

1.2.2 Impulse and Momentum

The motivation to define the quantities we call ***impulse*** and ***momentum*** comes directly from the dynamics equation:

$$\vec{F} \;=\; \frac{d\vec{p}}{dt} \;=\; \frac{d(m\vec{v})}{dt}\,;$$

$$\Rightarrow \quad \vec{F}dt \;=\; d(m\vec{v}).$$

The last expression above is $\vec{F} = m\vec{a}$ in differential form. What follows illustrates that it contains considerable information about the motion of the point mass m.

First consider the differential expression as it stands. It describes a relation between the force, the mass and the velocity over a differential time period dt. The product on the left side of the equation is pertinent to the motion and is given a special name: it is called the (differential) ***impulse*** delivered by the force to the mass over the time period dt. This impulse inherits the ***causal*** role which was assigned to the force. The reactive effect of the impulse on the motion of the mass is the right side of the equation – a (differential) change in the product $m\vec{v}$. This product is also an important parameter of the motion: it is called the ***momentum***.

---

[4] It will be seen that the quantities listed and their relations to each other are most easily approached from a differential viewpoint. It is interesting to note in this regard that calculus was unknown before the time of Newton. He devised it to describe the motions of masses. Also, calculus was discovered simultaneously and independently by a mathematician named Leibnitz. It appears that historically, science was ripe for its discovery.

In the cause-effect context, it can be said that the impulse delivered by the force ***causes*** a change in the momentum of the mass. The above equation quantifies this cause-effect relation. It is the differential form of the first and simplest of the three most basic laws of classical mechanics:

Impulse = Change of Momentum.

It should be noted that the analysis above indicates that the relation holds for every differential time period which is equivalent to saying that it is true instant by instant throughout the motion of the point mass.

To obtain the expression that applies to a finite time period (t_1, t_2) (as opposed to an infinitesimal period dt), the contributions from each differential time are added together (integrated):

$$\int_{t_1}^{t_2} \vec{F}\, dt \;=\; \int_{\vec{p}_1}^{\vec{p}_2} d\vec{p} \;\left(= \int_{m\vec{v}_1}^{m\vec{v}_2} d(m\vec{v})\right)$$

The integral on the left is called the ***impulse***, usually designated by an $\vec{I}$. It is referred to as the 'impulse delivered by the force $\vec{F}$, during the time interval $(t_1,\ t_2)$'. Note that the limits on the integrals must match: for example, $\vec{p}_1$ is the momentum at time t_1 and $\vec{p}_2$ is the momentum at time t_2. The result of integrating this last equation is

$$I \;=\; \vec{p}_2 - \vec{p}_1 \;(= mv_2 - mv_1)$$

This is the ***impulse-momentum law*** as it is usually presented. Stated in words,

> *the impulse delivered by a force over any time period equals the change in the momentum of the mass over that same time period.*

Note that in the integral definition of the impulse, $\vec{F}$ in the most general case can vary in both magnitude and direction during the course of the motion. Therefore the direct calculation of the integral requires vector addition which means that it must be evaluated for each component of $\vec{F}$ separately. However, the actual integral that defines the impulse is not used too often because the force is seldom known as a function of time: it is more often known as a function of position. To determine its time-dependence in these cases, one would have to know where the object was as a function of time. This means that the problem would have to be solved before the integral for the impulse could be set up.

It was stated earlier that Newton's first law (the law of inertia) was superfluous because it was an immediate consequence of the second (the dynamics law $F = ma$). That this is so can be shown by:

$$\vec{F}dt \quad = \quad d(m\vec{v});$$

but then

$$\left(\vec{F} = \vec{0}\right) \Rightarrow \left(d(m\vec{v}) = \vec{0}\right) \Rightarrow (m\vec{v} \text{ is constant})^5$$

The third parentheses in the logical string of statements above is just another way to state the law of inertia.

1.2.3 Work and Kinetic Energy

Unlike the previous result the work-energy law is not a vector relation: it is a scaler relation. Its mathematical derivation involves a kind of vector product called a dot product (or scaler product) which is a way of combining

---

[5] Statements made in the context of a logical string will be enclosed in parentheses.

two vectors to produce a scaler. It also involves the rules for differentiating this product. Although the derivation is fairly simple, it is too much of a diversion from the current subject matter to develop the mathematical machinery necessary to present it at this point. (It will be done later in *section* 3. 2. 7. 2.)

Instead, the relationship between work and kinetic energy will be derived using vector diagrams. Such diagrams are often useful in physics and therefore it is of some use to acquire the knack of understanding and producing them.[6]

Consider any arbitrary point along the path of a moving point mass. The force acting on the mass may be at any angle to (the line tangent to) the path. This force is resolved into two components, one in the direction of motion and one perpendicular to it. It will be seen in what follows that these two components have different effects on the motion. More specifically, it is asserted that:

1. *the component of force in the direction of motion (the tangential component) changes only the speed of the object;*

2. *the component of force perpendicular to the direction of motion (called the normal component) changes only the direction of motion.*

---

[6] A word about this knack: its essence is often hidden in the fact that any angles and distances which are differential in size cannot be represented as such: it would render the diagram incomprehensible. They must be drawn larger in order to see them. Consequently, as you use the diagram it must be kept in mind which quantities are differential even though they do not look that way. In doing this, it becomes easier to determine what small angle and/or small distance approximations can be used.

The quantitative version of the first of these is the work-energy relation to be derived in what follows; the second is associated with circular motion and will be discussed later in its proper context. The vector diagrams that follow will afford insight into both.

1.2.3.1   Proof of the Assertions

A point mass m is at an arbitrary point P along its path of motion. (Refer to *figure* $1.1\,a$.) At that instant a force $\vec{F}$ is acting on the body at an angle θ to its direction of motion. A short time Δt later, the mass is at a point Q traveling with the new velocity which results from the action of the force.

Figure $1.1.b$ shows the velocity vectors at P and Q with their tails referenced to the same point so they can be subtracted graphically to find $\Delta\vec{v}$. Note that the force is in the direction of the acceleration (since $\vec{F} = m\vec{a}$) and that it is (almost) in the direction of the change in the velocity $\Delta\vec{v}$. Therefore $\vec{F}$ is approximately parallel to $\Delta\vec{v}$ and it will become exactly parallel when $\Delta t \rightarrow 0$. Hence the angle at the top of the diagram is $(\theta - \Delta\phi)$. (In the limit all quantities prefixed with a $'\Delta'$ will approach 0.) Note also that the final velocity is divided into two pieces which are added together to find its length: the lower segment has length $|\vec{v}|\,cos(\Delta\phi)$ and the upper segment has length $|\Delta\vec{v}|\,cos(\theta)$.

We now calculate the change in *speed* of the mass. It is the magnitude of the final velocity of the mass minus the magnitude of its initial velocity. If Δs represents the change in speed, we have

$$\Delta s \;=\; |\vec{v} + \Delta\vec{v}| \;-\; |\vec{v}|$$

$$ds \;=\; \lim_{\Delta t \to 0}(|\vec{v}|\cos(\Delta\varphi) \;+\; |\Delta\vec{v}|\cos(\theta) \;-\; |\vec{v}|)$$

Using the approximation for the cosine of small angles $\left(cos(x) \approx 1 - \frac{x^2}{2}\right)$, and retaining only first order differentials, the first and third terms in the expression for **ds** cancel and the differential change in the speed becomes

$$ds \;=\; |d\vec{v}|cos(\theta)$$

This is the result we were after: the last equation contains enough information to prove both assertions.

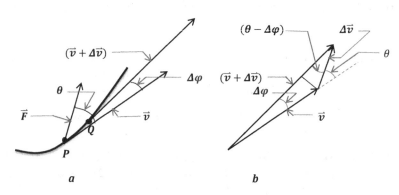

figure 1.1

Tangential and Normal Components of Force

We ask two questions:

1. under what conditions is the speed left unchanged? This is the case when only the direction is affected;

42

2. *under what conditions is the direction left unchanged? This is the case when only the speed is affected.*

In both instances, the case $|d\vec{v}| = 0$ is ignored since this corresponds to the case when the force is zero which is of no interest in this context.

To answer the first: the speed is constant when $ds = 0$. This occurs when

$$ds = |d\vec{v}| cos(\theta) = 0$$

$$\Rightarrow \quad \theta = \pi/2$$

$$\Rightarrow \quad \vec{F} \text{ is normal to the direction of motion.}$$

Therefore, the component of force normal to the direction of motion changes only the direction of motion without affecting the speed.

To answer the second: notice that 'no change in direction' implies that $\Delta\phi = 0$ which in turn implies that the triangle of *figure* 1.1.b collapses into a straight line. But then

$$\theta = 0$$

$$\Rightarrow \quad \vec{F} \text{ points in the direction of motion}$$

Therefore the component of force in the direction of motion changes only the speed without affecting the direction. We have:

1. *the normal component of force affects only the direction of motion;*
2. *the tangential component of force affects only the speed of the motion.*

The proof is complete.

In the general case, the force is neither tangential nor normal to the direction of motion but rather at some arbitrary angle $\boldsymbol{\theta}$ to it, as shown in $\boldsymbol{figure\ 1.1.a.}$ However it can always be replaced by its components in these directions. Indeed the foregoing analysis which associates the tangential component exclusively with speed changes and the normal component, exclusively with direction changes provides considerable motivation for doing so.

1.2.3.2 Derivation of the Work-Energy Relation

The content of the previous section motivates us to ask, "How does the tangential component of the force affect the speed of the body on which it acts?" The task of this section is to obtain a quantitative answer to this question. The result is the work-energy relation.

Once again we start with Newton's dynamics equation

$$\vec{F} \ = \ m\frac{d\vec{v}}{dt}$$

Since we are interested only in the tangential component of force, we single it out by multiplying both sides of the above equation by $\boldsymbol{cos(\theta)}$ where $\boldsymbol{\theta}$ is the angle between the force and the direction of motion:

$$\vec{F}\,cos(\theta) \ = \ m\,cos(\theta)\frac{d\vec{v}}{dt}$$

We know two things:

1. *the component of force on the left is related somehow to the speed of the object;*

2. *the speed is $\frac{ds}{dt}$ where s is distance along the path of motion.*

We attempt to involve the speed of the body. Note that the $\vec{v}$ on the right is not suitable for this purpose because it is not the speed but the velocity which exhibits a change in direction as well as a change in magnitude. Since there is already a *dt* in the 'denominator' of the derivative, we involve the speed by multiplying both sides of the equation by *ds:*

$$\vec{F}\cos(\theta)\,ds \;=\; m\cos(\theta)\,d\vec{v}\frac{ds}{dt}.$$

Consider the right side of this equation: $\frac{ds}{dt}$ is the speed and $cos(\theta)\,d\vec{v}$ is the tangential component of the velocity change. Reference to the previous section shows that the tangential component of the velocity change is the speed change (the normal component of the velocity change pertains only to a directional change).

We now have an equation which involves only the tangential component of force and the speed of the object:

$$\vec{F}_s\,ds \;=\; m|\vec{v}|d|\vec{v}|,$$

where the subscript *s* indicates the component of $\vec{F}$ in the direction of motion. Since the speed is a scaler, the right side can be integrated and the result follows:

$$\int_P^Q \vec{F}_s\,ds \;=\; \int_{|\vec{v}_1|}^{|\vec{v}_2|} m\,|\vec{v}|\,d|\vec{v}|$$

$$\int_P^Q \vec{F}_s\,ds \;=\; \frac{1}{2}\,m(|\vec{v}_2|^2 - |\vec{v}_1|^2)$$

As before the quantities involved in this equation occur often because they follow so closely from $\vec{F} = m\vec{a}$. They are given special names: the integral on the left is called the **work** done by the force on the body; the quantity

$$\frac{1}{2}m|\vec{v}|^2$$

on the right is called the **kinetic energy** of the body. Stated in words, the above relation says

Work   =   Change in Kinetic Energy

Notice that the work defined by the integral on the left above involves only the tangential component of force and according to the results of the analysis of the previous vector diagram, this component affects only the speed of the mass. The work-energy relation above is the precise quantitative statement which describes the effect of this component on the speed on the motion.

It often happens that the work integral appears to be a rabbit pulled out of a hat because usually in a first course in mechanics there is insufficient background in vector calculus to describe adequately how it follows from the dynamics law. It is hoped that the previous remarks and analysis serve to remove some of the mystery.

By the way, the work integral is no hat-dwelling rabbit; it is more like a bottle-dwelling genie ready to grant your every wish. It provides the basis for the concepts of both potential and kinetic energy. Keep in mind that the work-energy relation is scaler and much less tedious to manage than the more usual vector relations in mechanics.

Note that the work-energy relation is completely general:

> *the work done by a force on a mass as it moves from any arbitrary point P to any other arbitrary point Q is equal to the change in the kinetic energy of the mass between the same two points.*

(Remember that the component of force normal to the direction of motion does not alter the speed and therefore has no effect on the kinetic energy.)

1.2 CONSERVATIVE FORCES AND POTENTIAL ENERGY

1.3.1 Conservative and Non-Conservative Forces

Up to this point little attention has been given to the nature of the particular forces acting on a body. Now it is time to focus on them to see if anything can be said about them in general. The doorway into this investigation is the work integral:

$$W = \int F_s \, ds.$$

In the most general cases, the evaluation of the work integral comes with a catch 22: in order to find the path of motion, the forces must be known; but when the forces are a function of position, the forces are not known until the path of motion is known. Many of the methods used to approach this issue involve educated guesses combined with successive approximations. These are long tedious processes and the solutions they offer are for the most part 'ad hoc' rather than general.

There are cases however where the work integral lends itself to a more satisfying analysis. Such cases are considered in what follows.

We make a wild conjecture (not without foresight). Suppose a mass moves from point A to point B along some given path and we calculate the work done by the forces in play. Then suppose it moves from A to B along a different path and again we find the work done by the forces... then a third, fourth and fifth time and so on into the night choosing a different path from **A** to **B** each time. The wild conjecture is not that we would do that (although that's pretty wild in itself) but rather that *we got the same answer every time*. In other words, we are determining that the particular force field acting on the body is such that

> ***the work done in going from one arbitrary point to another does not depend on the path over which it moves but only on the endpoints A and B of the motion.***

This would seem very unlikely and indeed among the infinitude of mathematically possible forces, there are 'few' that would fit this conjecture. But this is physics, not mathematics and the fact is that nature seems to favor this type of force field. Among the important forces that do fit this conjecture are the local gravity force field ($F = mg$) or any constant force field for that matter, the universal gravity field together with the electrostatic field (the force in each of these two cases varies as the inverse square of the distance from the source of the field) and restoring forces (springs which give rise to oscillatory motion). Generally, all force fields which fit the conjecture are called **conservative** force fields.

All forces which are not conservative are called **non-conservative** For example, **friction** is a non-conservative force. Consider a block that is pushed from point **A** to point **B** on a non-frictionless table top: if it is pushed in a straight line from **A** to **B**, a certain amount of work is done to overcome the friction; if it is pushed all over the table along some complex path and finally to **B**, the work done against the friction force is greater because the path was longer. Hence friction forces are not independent of the path of motion and are therefore non-conservative.

1.3.2   The Mathematics of Conservative Forces

Refer to *figure* 1.2 which shows a force *F* acting on a body at a certain point *P* as it moves along its path of motion. Let *θ* be the angle between the force a

nd the path of motion and consider the work integral,

$$W \;=\; \int |\vec{F}| \cos(\theta) \, ds \, .$$

The complications in evaluating the integral are the result of the following facts concerning the integrand:

1. *$\vec{F}$ is in general not constant. It varies from point to point in both magnitude and direction (and sometimes with time, but we are not considering that case here). It is, however defined at all points in space and is independent of the path a body may traverse through it;*

2. *The differential path length ds is*

$$ds \;=\; \sqrt{(dx)^2 \;+\; (dy)^2}$$
$$=\; \sqrt{1 \;+\; \left(\frac{dy}{dx}\right)^2} \; dx$$

   and is highly dependent on the path of motion since $\frac{dy}{dx}$ is the slope of the path;

3. *the angle θ is also dependent on the path since the tangent to the path of motion is one side of the angle.*

Consequently, the integral cannot be evaluated in any general sense and will in most cases take on different values for different paths.

$$W = \int_P^Q \vec{F}_s \, ds$$

$$ds = \sqrt{(dx)^2 + (dy)^2} = \sqrt{1 + \left(\frac{dy}{dx}\right)^2}\, dx$$

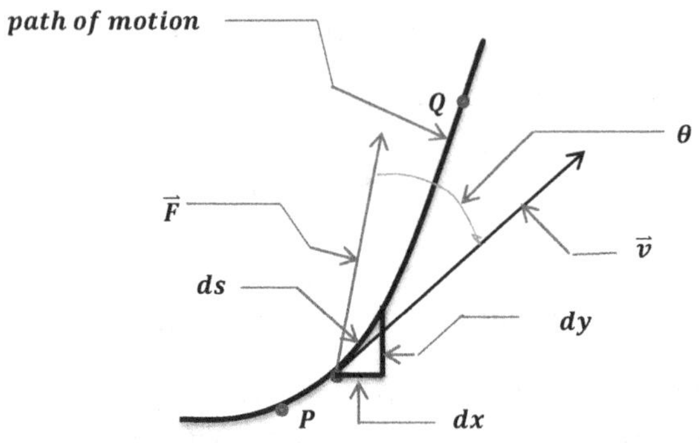

figure 1.2

The Work Integral

We are interested in those cases where this situation just does not exist. We have defined conservative forces as those for which the work done in going from A to B (A and B arbitrary) is independent of the path. The implication is that the work integral can be evaluated without any knowledge whatsoever concerning the path of motion.

The fact that the value of the work integral is independent of the path of motion carries with it an important implication. The following argument will clarify it.

Let A be a fixed point and let U_P be the value of the work integral when the body moves from point A to point P. If A is kept fixed, and P is allowed to vary, U_P can be written as a function of P:

$$U_P = U(P).$$

Suppose this is done again but this time the varying point is labeled Q. Then

$$U_Q = U(Q)$$

Now if the body moves from A to some point Q going through point P along the way, we must have

$$U(Q) = U(P) + U_{PQ}$$

where the last term is the work to go from P to Q (which has nothing to do with the reference point A).

This last equation is true for arbitrary points P and Q *if and only if* the work is independent of the path of motion. Hence,

$$U_{PQ} = U(Q) - U(P).$$

The last expression implies that the work in going from any arbitrary point P to another arbitrary point Q can be written as some function U evaluated at Q minus its value at P regardless of the reference point A. This is a relation that is peculiar to the situation when the work integral is independent of the path of motion. If the work integral is not independent of the path of motion, no such function exists. Therefore *'independent of the path of motion'* implies the *existence* of a function U such that

$$W = \int_P^Q |\vec{F}| \cos(\theta) \, ds = U(Q) - U(P)$$

Since the integrand is a scaler, so is U Also, U is a function of position and it is essentially the work integral already evaluated.

The above equation is similar to the conclusion of the fundamental theorem of integral calculus (see *section* $0.3.6$). The temptation is to say, *"if U is the integral of $\vec{F}$ then $\vec{F}$ is the derivative of U."* The only problem is that $\vec{F}$ is a vector and U is a scaler. The statement is therefore a little loose: that it is essentially correct however, is the subject of the next section.

1.3.3   Partial Derivatives and Potential Energy

In the interest of finding the relation between U and $\vec{F}$ consider the following. Suppose $\vec{F}$ is a conservative force field and let dW be the work done by $\vec{F}$ along some differential path. Since the work is independent of the path, dW can be calculated using *any path*. We are therefore free to choose a simple one. First move a distance dx in the x-direction. The contribution to the work is $F_x dx$, (the component of force in the direction of motion times the distance). If we then move similarly in the y- and z-directions, the total differential work delivered by the force is the sum of the three contributions

$$dW \;=\; F_x\,dx \;+\; F_y\,dy \;+\; F_z\,dz$$

or, since W and U are essentially the same thing

$$dU \;=\; F_x\,dx \;+\; F_y\,dy \;+\; F_z\,dz$$

But the differential change, dU can be expressed in another way – in terms of its own changes. As we moved a distance dx along the x-axis, (note that y and z are constant during this process), the contribution to the change in U can be represented by

$$\frac{\partial U}{\partial x} \, dx$$

or, in words, the change of U per unit change in x times the change in x. Adding the contributions from each direction, the total change in U is

$$dU \;=\; \frac{\partial U}{\partial x} \, dx \;+\; \frac{\partial U}{\partial y} \, dy \;+\; \frac{\partial U}{\partial z} \, dz$$

The use of the Greek *script* delta's in the above derivatives indicates for example that y and z are treated as constants when taking the derivative with respect to x. (The contribution to dU from the first term on the right was computed from movement in the x-direction: y and z were constant during this process; the script delta's are used in the derivative $\frac{\partial U}{\partial x}$ to reflect this fact. Similar statements apply to each of the other two terms.) These are called *partial derivatives* and the previous statement determines that the rule for finding a partial derivative is

> *take the derivative with respect to the variable indicated treating all other (independent) variables as if they were constants.*

Finally, it is more convenient to express things in terms of the negative of the function U. Hence, we define a new function: let $V = -U$. This does not affect the physics at all: it simply makes some pertinent quantities appear with '+' instead of '-' signs.

The final result of all this is seen by comparing the expressions for dW and dU and expressing the results in terms of V. These relations are true for **any** point differentially displaced from P. It follows that dx, dy and *dz* can be chosen independently. Hence each of their coefficients can be identified separately and we have:

$$\vec{F}_x = \frac{\partial U}{\partial x} = -\frac{\partial V}{\partial x}$$

$$\vec{F}_y = \frac{\partial U}{\partial y} = -\frac{\partial V}{\partial y}$$

$$\vec{F}_z = \frac{\partial U}{\partial z} = -\frac{\partial V}{\partial z}$$

Thus each of the components of the force is an appropriate (partial) derivative of *V*. It is in this sense that $\vec{F}$ is a derivative of *V* and the loose statement made above is explained. (See the last paragraph of ***section* 1.3.2**.)

When such a function *V* exists, it is called the ***potential energy*** and the associated force field is said to be ***derivable from a potential***. When this is the case, there is no need to deal with the work integral at all: all the quantities of interest are easily found from the potential energy function *V*.

To reiterate what precedes, the difficulties encountered when one tries to evaluate the work integral were discussed. The 'wild conjecture' was made that there exist force fields for which the work done in traveling from any arbitrary point to any other was independent of the path, and it was found that in such cases a scaler function of position could be defined which is essentially the work integral already evaluated.

In continuing this line of investigation, we will consider only those force fields which depend on the position. If more complex situations can be treated at all, they require elaborations of these techniques. The point is always the same however: the motion of a body under the action of some force is greatly simplified when it can be determined that the work integral is independent of the path of motion.

As an example, consider the local gravity force as representative of any constant force field (one which is the same everywhere in the space). A body in this field is subject to a force equal to its weight ($\boldsymbol{F} = \boldsymbol{mg}$) pointed toward the surface of the earth. Refer to *figure* 1.3 which shows a mass sliding down a frictionless surface along an unknown path near the surface of the earth. The forces acting on the mass are gravity and the contact force $\overline{N}$. (Note that a frictionless surface can sustain only a normal force.)

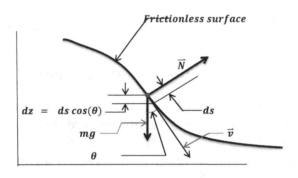

figure 1.3

Local Gravity: A Conservative Force Field

The work integral is

$$W \;=\; \int \vec{F} \cos(\theta)\, ds\,,$$

where we have no information concerning the angle θ or the differential path length $\boldsymbol{ds}$. The integral can be evaluated anyway because no matter what the shape of the path, the product $cos(\theta)\, ds$ is dz, the differential change in height. The work integral then is

$$W \;=\; -\int_{z_1}^{z_2} mg\, dz$$

$$W \;=\; -mg(z_2 - z_1)$$

The integral was evaluated without any knowledge of the path and therefore from what was said above, it can be concluded that the local gravitational field (and, in fact any constant force field) is conservative. In this instance, the work depended on only the initial and final values of z. (Given any two endpoints A and B of the motion, these values of z are known and hence so is the work done by the force.)

1.3.4 The Definition of Potential Energy

The implication of the previous two sections is that whenever conservative forces are acting, there is no necessity to deal with the work integral: it is possible to find its indefinite integral once and for all and calculate the work in going from one point to another by taking the difference of the (integrated) function evaluated at the endpoints.

When this is done, the integrated function is called the *potential energy* and the force is said to be *conservative*; it is also said that the force is *derivable from a potential*.

The potential energy function for the local gravitational field is

$$V \;=\; mgz,$$

where z is the height above the surface of the earth. Notice that

$$F_z \;=\; -mg \;=\; -\frac{\partial V}{\partial z}$$

in keeping with the analysis above.

Actually, we have ignored the constant of integration which comes from the evaluation of the indefinite work integral. Note that only **differences** in potential energy are pertinent (at least in this context) and the constant always cancels itself out when you take the difference. The point of zero potential energy can be set wherever it is convenient. In this case it was taken at the surface of the earth where $z = 0$.

The potential energy of a conservative force field is defined as

$$V \;=\; -\int F_s\, ds \;=\; -\int |\vec{F}|\cos(\theta)\, ds \;:$$

but it can be defined only when the integrand can be expressed independently of the path.

Since the potential energy is just the integrated version of the work integral, it can be substituted in the work-energy relation:

$$W \;=\; \int_A^B |\vec{F}|\cos(\theta)ds \;=\; -V(B) + V(A)$$
$$=\; T(B) - T(A),$$

where T represents the kinetic energy of the body. Transposing terms, we have

$$T(A) + V(A) \;=\; T(B) + V(B)$$

which says that the sum of the kinetic and potential energies is the same at A as it is at B. But A and B are arbitrary. It follows that the sum of the two energies is constant throughout the motion. If **H** is the total energy, then

$$H \;=\; T \;+\; V$$

$$H \;=\; constant\ throughout$$
$$the\ motion.$$

Whenever a quantity is constant throughout the motion, it is said to be **conserved**. It has just been shown that

> **whenever all the forces acting on a system are conservative, the total energy of the system is conserved.**

(That is why forces that are derivable from a potential are dubbed *'conservative'.*)

1.3.5   Some Conservative Force Fields

1.3.5.1   Constant Force Fields

A constant force field is one in which the force is the same (in both magnitude and direction) no matter where you are in the field. All such fields are essentially the same as the local gravitational field which was already shown to be conservative (refer to *section* **1.3.2**).

Let **r** be distance measured in the direction of the force of magnitude **k** in a constant force field. Then the potential energy is

$$V \;=\; -\int k\cos(\theta)\,ds \;=\; -k\int dr$$

$$V \;=\; -kr \;+\; constant$$

where **θ** is the angle between the force vector and the direction of motion.

1.3.5.2 One-Dimensional Force Fields

Suppose that a body can move in only one dimension (along the x-axis, say) and that the force acting on it is a function only of its position x along the axis. It follows that the component of force in the x-direction also depends only on x. Then the integrand of the work integral depends only on x. Furthermore the differential path length is always dx. These facts imply that the integral can always be evaluated to find the potential energy function.

Thus spring forces are conservative because they act in one dimension and the force depends only on the displacement from equilibrium. This is a specific one-dimensional force singled out because it is important: it gives rise to simple harmoinic motion which governs all wave analyses.

A spring force is defined by the equation

$$F_x = -kx.$$

The quantity k is called the spring constant which tells you how hard the spring pushes or pulls on the mass: the minus sign means that the force tends to bring the body back to its equilibrium at $x = 0$. The magnitude of the force is proportional to the displacement x. For these reasons, a spring force is referred to as a *linear (simple) restoring force*: *'linear'* (*'simple'*) because it is directly proportional to the displacement; *'restoring'* because it tends to bring the system back to its equilibrium.

Let k be the spring constant for a given spring. Then the potential energy is

$$V \;=\; -\int |F| \cos(\theta)\, ds \;\cdot\; = \;\int k\, x dx$$

$$V \;=\; \frac{1}{2}\, k\, x^2$$

In one-dimensional motion, since all the geometric problems associated with the angle θ and the differential path length are automatically eliminated, there is a temptation to say that friction becomes a conservative force. This is clearly an erroneous conclusion because friction always opposes the motion: when the mass moves past a point x to the right, the friction force is to the left; when the mass moves past the same point x to the left, the friction force is to the right. Therefore the friction force is **not** a function only of the position x; it depends also on the direction of motion. (It could depend also on the velocity, etc.)

1.3.5.3 Radially Directed Fields

A radially directed force is one which always points toward (or away from) a centrally located point. In general its magnitude can depend on all three coördinates (r, φ, θ).

From the diagram in ***figure*** **1. 4** it is seen that

$$\cos(\alpha)\, ds \;=\; dr$$

where α is the angle between the direction of motion and the radially directed force.

The work integral

$$W \;=\; \int F(r, \varphi, \theta)\, \cos(\alpha)\, ds \;=\; \int F(r, \varphi, \theta) dr$$

60

cannot be evaluated in general owing to the dependence of the integrand on the angles φ and θ which are different for different paths between two given endpoints. Hence this is in general a non-conservative field. If however, the force depends *only on the distance r* from the central point (i.e. it is independent of the angular displacements φ and θ) then

$$W \;=\; \int F(r)dr$$

which can be integrated without knowing the path of motion. Hence if the magnitude of the force in a centrally directed field depends only on the distance r from the central point, the force field is conservative.

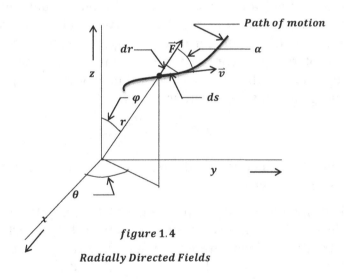

figure 1.4

Radially Directed Fields

The gravitational field of the earth, for example is directed towards its center (of mass) and attracts a body of mass m with a force

$$F = -G\frac{Mm}{r^2},$$

where G is the universal gravitational constant, M is the mass of the earth and r is the distance between the center of the earth and the mass m. It satisfies the above requirements and is therefore a conservative force field.

The potential energy for this field is

$$V = -\int F(r) \cos(\alpha)ds = -\int F(r)dr$$
$$V = GMm\int dr/r^2$$
$$V = -\frac{GMm}{r}.$$

The electrostatic field is also a centrally directed inverse square field and hence is conservative. It describes the force of attraction or repulsion between electrical charges and has the same form as the gravitational force field except that the constant is different and the masses are replaced by the charges (with their signs). The potential energy for this force field is

$$V = -\frac{K q_1 q_2}{r},$$

where K is a constant, the $q's$ are the charges and r is the distance between them.

The inverse square dependence of the gravitational and electro-static fields is not a new relation arbitrarily introduced. These are experimental results: in the case of the electro-static field, it is the direct result of experimental measurement; in the case of the gravitational field, it was determined from Kepler's laws (see *section* 6.1.4) and later by direct result of laboratory experiments.

3.5.4 The Vector Sum of Conservative Force Fields

Suppose for a moment that two charged masses are connected by a spring and each is connected by other springs to the ceiling of a room that is spinning as it orbits in the gravitational field of the earth. Then someone says, "Describe the motion": a truly complex problem.

After punching him in the nose for ever thinking that yo **definite integral** u should solve such a problem, you might go about the business of drawing force diagrams and resolving all the forces into components, etc. ad infinitum.

But a better stab can be made at this problem. Notice that all the forces acting are conservative. So consider this: if one were to compute the work done by each of these forces as the body moves through some distance, he would end up with a bunch of work integrals added together, and after integrating them he would have the sum of their associated potential energies. There's the clue! When a number of fields are acting simultaneously, the vector forces are added together (this is a vector addition). But their potentials also add together (this is a scaler addition). This is so because the integral of the sum equals the sum of the integrals a fact which is mathematically trivial but has far-reaching implications in potential theory. It means that the sum of conservative force fields is also conservative and that the potential energy of their sum is the scaler sum of their individual potential energies.

Thus if there are three electrical charges, the first moving in the field of the other two which are in fixed locations, we have a situation where the field in which the first is moving has a fairly complicated configuration and is not radially directed since the other two are at different locations. We are assured however that the resulting field is conservative and that its potential energy is simply the sum of the potential energies of the component fields.

We have

> **any sum of conservative force fields is also conservative and the potential energy of the sum is the sum of the potential energies of the individual component fields.**

The initial 'wild conjecture' that started all this has brought us a long way. We see that the analyses of a great number of physical situations can be considerably simplified by the use of potential energy functions.

The theory of potentials together with its intricacies and extensions has been studied extensively and is the sole subject of many books.

1.4   SUMMARY

All the basic quantities associated with linear motion have been presented and defined and their interrelations have been shown to be the direct result of Newton's law of dynamics $\vec{F} = m\vec{a}$. The following is a recapitulation of these definitions and results in a condensed form.

The quantities defined were:

name	Definition	Symbol
Momentum	$m\vec{v}$	$\vec{p}$
Impulse	$\int \overrightarrow{Fdt}$	$I \text{ or } J$
Work	$\int F_s ds$	W
Kinetic energy	$\frac{1}{2}mv^2$	T
Potential Energy	$-\int F_s ds$	$V \left(F_s = -\frac{\delta V}{\delta s}\right)$
Total Energy	$T + V$	H

The last two rows pertain to conservative systems. (Otherwise the potential energy cannot be defined.)

Laws following directly from $\vec{F} = m\vec{a}$:

1. Impulse   =   Change in Momentum

$$\int_{t_1}^{t_2} \vec{F}\, dt \quad = \quad m\vec{v}_2 \; - \; m\vec{v}_1 .$$

This law always holds and is useful when the impulse is given or when $\vec{F}$ is known as a function of time during the motion.

2. Work   =   Change in Kinetic Energy

$$\int_A^B F_s\, ds \quad = \quad \tfrac{1}{2}mv_B^2 - \tfrac{1}{2}mv_A^2$$

This law always holds and is useful when $\vec{F}$ is known as a function of position and the geometry is manageable.

3. *Conservation of Total Energy*

$$H \;=\; T(A) + V(A) \;=\; T(B) + V(B)$$

This law holds when all the forces are conservative and their associated work integrals have been replaced with the corresponding potential energies.

The force acting on a mass in motion was considered. It was found that the component of force normal to the path of motion changed only the direction of motion while the component in the direction of motion changed only the speed. This result motivated the decomposition of the force vector into two components, one parallel to and one normal to the path of motion. The idea was to quantify the effect of the force on the speed by singling out the component in the direction of motion. The result was the work-energy relation.

Finally, there was a discussion of the work integral to see what could be said about it in general. It was found that in many instances, it could be evaluated without any knowledge of the path of motion, this being possible if and only if all the forces acting were conservative. When this is the case a potential energy function can be defined (which is the work integral already evaluated) and the total energy, potential plus kinetic is constant throughout the motion, i.e. total energy is conserved.

CHAPTER II

The Extension of Newton's Laws to All Material Systems

2.1 PREVIEW

2.1.1 Purpose

This preview offers a few generalized statements describing what was done in the previous chapter and serves as an outline for the current one.

In the previous chapter, the idealization was made that the mass of an object was concentrated at a single point. This was done to avoid issues of the shape and extent of the body as well as the necessity to consider any rotational motion it might acquire. It was assumed only that Newton's laws were valid for point masses. His equations were used to define the quantities ever-present in mechanics and to derive the dynamic relations between them. Although this allowed for a preliminary look at the most basic principles of Newtonian mechanics in their simplest form, the assumption that each mass was concentrated at a point is admittedly unrealistic.

It is the purpose of this chapter to show that this assumption tolerated till now, implies that Newton's laws hold also for any arbitrary extended (real) body whether it be gas, liquid or solid. No additional assumptions will be necessary to obtain this result. It will justify the application of the laws previously derived for point masses to any arbitrary extended system.

2.1.2 Moments and the See-Saw

In order to accomplish the task at hand, it is necessary to have a concept of the center of mass of a body. Nearly everyone has at least the vague sense that the center of mass of a body or of any collection of masses is the point around which all the mass is equally distributed. An analysis is presented which relates this concept to the balancing of a see-saw: it will establish mass moments (mass times distance from the fulcrum) as the quantities pertinent to balancing a given mass distribution. Generally speaking, moments are quantities which pertain to rotation. The precise sense in which the mass is 'equally distributed' around the center of mass is contained in the

> ***Definition: the Center of Mass of a body is that point around which all the mass moments add to zero.***

2.1.3 Equilibrium and Potential Energy

This section contains a discussion of equilibrium. Three types are considered, ***stable***, ***unstable*** and a third kind referred to here as ***extensive*** because it extends throughout some local region. The material on equilibrium is then viewed from the perspective of potentials where the conditions for equilibrium discussed previously are translated into conditions on the potential energy. The purpose is to familiarize the reader with some manipulations and applications of potentials which generally affect a great simplification of problems involving conservative forces.

2.1.4 Center of Mass

This section contains the derivation of the formulas for determining the position, velocity and acceleration of the center of mass of an arbitrary collection of extended masses. These formulas are subsequently used to extend the realm of application of Newton's laws to any system of material bodies.

2.1.5 Arbitrary Collections of Point Masses

A system acted upon by arbitrary external forces and containing an arbitrary collection of point masses exerting arbitrary forces on each other is considered. It is shown that regardless of these internal forces, the *CM* (center of mass) of the system moves according to Newton's prescriptions, i.e.

$$\vec{F} = M\vec{A},$$

where $\vec{F}$ is the (vector) sum of all the **external** forces, M is the total mass of the collection and $\vec{A}$ is the acceleration of the center of mass.

It should be noted that the point masses considered may be individual molecules or differential-size chunks of matter; the internal forces may be pressures moving a liquid or compressing/rarefying a gas, intermolecular forces holding a solid together, etc. The situation considered is completely general and the implication of the result is that Newton's laws apply to every material body and more generally, to any system of masses.

2.2   THE SECRET OF THE SEE-SAW

2.2.1 Two Masses on a See-Saw

Consider the situation as shown in ***figure* 2. 1**. There is a fulcrum with a cross-bar (assumed massless) with a mass **M** on one side a distance **R** from the fulcrum and a mass **m** on the other side at a distance **r** from the fulcrum. The see-saw is shown in an arbitrary initial position and also in a second position such that the height **H** of the mass **M** from the horizontal reference changes by **ΔH**. Use will be made of the fact that the left and right sides of the figure are geometrically similar. Let the proportionality constant be **k (r = kR)**.

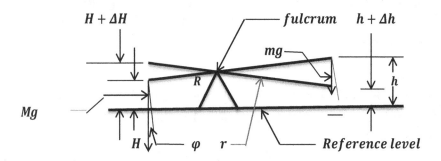

figure 2. 1

Balancing a See Saw

First, resolve the forces **mg** and **Mg** into components parallel and perpendicular to the board and note that only the perpendicular components tend to rotate the board around the fulcrum (the parallel components tend to slide the board across the fulcrum). Note also that the force mg tends to rotate the system clockwise and the force Mg tends to rotate it counter-clockwise. We seek to find the equilibrium condition for the system. This will determine the mathematical expression which quantifies the tendency of the forces to rotate the system.

Suppose that initially, the system is balanced so that it is at rest. Since no part of the system is accelerating, the net force acting on it must be zero. It follows that if it is displaced differentially, the work done by the force of gravity must be zero because the net force acting at equilibrium is zero. The displacement must be differential in size however because after a slight displacement, it is not known whether or not the system is still in equilibrium. (It actually is but this will be part of the result and cannot be depended upon at this point.)

The following sequence is self-explanatory:

$$W \ = \ \int F_z \, dz$$

$$0 \ = \ -Mg \cos(\varphi)\Delta H \ - \ mg\cos(\varphi)\Delta h$$

$$0 \ = \ -Mg \cos(\varphi)\Delta H \ + \ mgk \cos(\varphi)\Delta H$$

After canceling and using the relation $r \ = \ kR$, the condition for balance is found to be

$$M \ = \ m\frac{r}{R}$$

$$MR \ = \ mr.$$

The condition for balance is independent of the angle, φ. This indicates that when the condition for balance is satisfied, the see-saw remains in a state of equilibrium regardless of the angle through which it is rotated.

The third equation in the above sequence indicates the quantities that tend to rotate the board around the fulcrum:

$$RMg \cos(\varphi) \longrightarrow \textit{tends to rotate the board counter-clockwise} ;$$

$$rmg \cos(\varphi) \longrightarrow \textit{tends to rotate the board clockwise} .$$

In each case, it is the component of force normal to the board times its distance from the fulcrum. (note that the component parallel to the board tends to slide the board across the fulcrum.) Therefore, the secret of the see-saw is:

> *the quantity that tends to rotate a system around a given point is the normal component of force multiplied by its distance from the point, or equivalently, the magnitude of the force times its perpendicular distance to the point.*

Exercise: Show that these two definitions are equivalent.

Exercise: Fill in the missing algebra in the above analysis.

If P is any point and M, a mass at a distance R from P, then the product MR is called the moment of mass around P. The distance R is called the moment arm. More generally, any scaler quantity, Q, which is endowed with a location, has a moment around P which is similarly defined.

> *all moments of scaler quantities will be defined as the product of the scaler quantity and the moment arm;*

> *all moments of vector quantities will be defined as the product of the vector component normal to the moment arm and the moment arm.*

The previous discussion establishes these definitions as *feasible* tools for the analysis of rotational motion. It will be seen later (when circular motion is considered in detail) that the quantity which tends to rotate a system around some given point is the *torque* which is the *moment of force* around that point.

2.3   EQUILIBRIUM AND POTENTIALS

2.3.1   The General Conditions for Equilibrium

Equilibrium is a state of balance. It comes in two colors: static and dynamic. Dynamic equilibrium is a state of balance in a system whose parts are moving like the solar system where everyone stays in his own orbit, takes care not to bunk into his neighbor and goes about the business of planning his next annual (a duration that varies from planet to planet) excursion around the sun. We leave this fascinating subject now because...

... we wish to make some observations about static equilibrium. This is a state of balance in a system whose parts are stationary. It is easy to presume that a system in this state will remain so until some external influence disrupts the balance. This is not always so. To be correct, one must consider three two types of static equilibrium, stable, unstable and extensive (neutral).

Consider a cone placed base-side-down on a table. It is obvious that it is in a state of balance and will remain so. Also, if displaced differentially by lifting its base slightly on one side, it would return to its previous state upon being released. This is called *stable static equilibrium.*

It is also obvious from symmetry that the cone is balanced when we attempt to place it point-side-down on the table.

But no attempt to get it to remain in this position is successful: it always topples. Furthermore, if it were possible to attain this delicate balance, any differential displacement would send it on further excursions from its initial balance. This is called ***unstable static equilibrium.***

Now, consider a marble on a flat table. It is in a state of equilibrium and remains so. If displaced slightly, however, it tends neither to return to its initial position nor to move further from it; it would be in a new equilibrium position. This kind of equilibrium is referred to here as ***extensive.*** The word is meant to imply that the equilibrium positions extend to a region containing the initial position as an internal point. (This type of equilibrium is also called ***neutral equilibrium.***)

What criterion determines whether or not a system is in a state of static equilibrium?... and given that it is, what are the criteria which determine whether the equilibrium is stable, unstable or extensive? We will answer these questions specifically for conservative systems but first, the more general descriptive answers will be given.

In what follows, consider a three-dimensional, many-body system.

A system is in a state of ***static equilibrium*** if and only if the net force acting on each of its components is zero: for if any component experiences a non-zero net force, $\vec{F} = m\vec{a}$ dictates that it will accelerate and the system would no longer be static.

To determine the type of equilibrium, consider each component of the system differentially displaced from the initial equilibrium. ***Stable equilibrium*** requires that for all possible configurations of differential displacement, each component experience a force that tends to return it to the equilibrium position. The equilibrium is **unstable** if for some such differential displacement, at least one component experiences a force which tends to move it farther away from its initial position. The equilibrium is **extensive** if it is not unstable and at least one component is not subject to any force at all when differentially displaced.

These are simple ideas which suggest a method for testing systems for stability: it is called he method of **virtual displacements.** The method was employed in the above definitions but needs some explanation. The displacements employed in this context have nothing to do with the dynamics of the system. They are a mathematical trick: we are saying that we want to check the system out (in this context, to determine the type of equilibrium) and that we are going to do so by answering the question, "What happens if the system is differentially close to the equilibrium state being tested?" So we are supposing all parts of the system to be displaced in an arbitrary way to see if it returns to equilibrium, meanders farther from it or remains in its displaced configuration. These are called virtual displacements to distinguish them from real displacement configurations to which the system may migrate as a result of the dynamics of its movements. These real displacements answer the question, "How will the system move if it is perturbed in some particular way?" The difference is in the appearance of time in the equations: real displacements occur in real time which appears explicitly in their description; virtual displacements are supposed configurations and do not involve time.

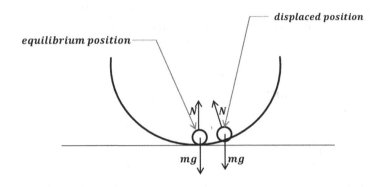

equilibrium position

displaced position

figure 2.2.*a*

Stable Equilibrium

Figure 2.2.*a* is a sketch of a hemispherical bowl with a marble in it. The equilibrium is obviously stable but let's check it anyway. In the equilibrium position, the normal force *N* exactly balances the weight *mg*. (Frictionless contact supports **only** a normal force: there can be no tangential component.) When the marble is slightly displaced, the normal force acquires a component which tends to bring it back to its equilibrium position. This is true no matter what the direction of the displacement. It is therefore concluded that the equilibrium is stable.

Now consider the sketch shown in *figure* 2.2.*b*. the bowl is turned over with the marble placed on the top. Again, the weight mg is exactly balanced by the normal force N. Since the net force is zero, the marble is in a state of equilibrium. But when it is differentially displaced, the normal force acquires a component which tends to displace it even further from equilibrium. It can therefore be concluded that the equilibrium is unstable.

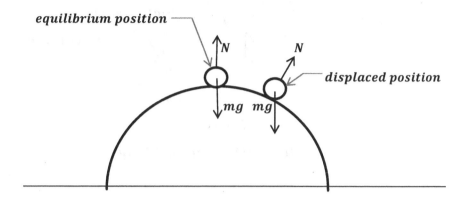

figure 2.2.b

Unstable Equilibrium

Now consider the sketch shown in *figure 2.2.b*. the bowl is turned over with the marble placed on the top. Again, the weight *mg* is exactly balanced by the normal force *N*. Since the net force is zero, the marble is in a state of equilibrium. But when it is differentially displaced, the normal force acquires a component which tends to displace it even further from equilibrium. It can therefore be concluded that the equilibrium is unstable.

Figure 2.2.c shows the marble in equilibrium on a flat frictionless table top. After displacement it is still in equilibrium tending neither to return nor displace further. The equilibrium is extensive.

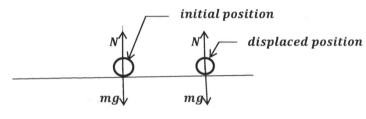

figure 2.2.c

Extensive (Neutral) Equilibrium

2.3.2 Equilibrium in Conservative Systems

In what precedes, equilibrium was identified with the condition that there is no net force. The type of equilibrium, stable, unstable or extensive was then determined by investigating the forces when the system was infinitesimally close to equilibrium. We proceed now to translate these ideas into the language of potential energy.

Recall that potential energy was defined as the negative of the work integral

$$V \;=\; -\int F_s \, ds \,,$$

where the subscript s indicates the component of $\vec{F}$ in the direction of motion and ds is a differential distance along the path of motion. The above equation indicates that V is the integral of the force $\vec{F}$ in some sense: we would therefore expect that the force is the derivative of the potential in some sense and this is indeed true but there's a...

Mathematical Glitch. In elementary calculus, one could wander back and forth along the x-axis and note the changes in some function $f(x)$ and in so doing define the derivative $\frac{df(x)}{dx}$ as the change in $f(x)$ per unit change in x. But now the constraint to move in only one direction is no longer there: we are free to meander in an infinite number of directions and note the changes in the function V (which is in general, a function of x, y and z) as we do so. Now this could get a little unruly because there is such a pluricity of directions in which we could choose to wander and each choice results in a different value for the change in V. Hence V generally has a different derivative for each choice of direction.

(This material was discussed earlier in a slightly different way.)

Take a closer look. Consider the above integral and suppose for the moment that s is some given direction. Suppose further that we are at a point P and seek to find the differential change in V when we move a distance Δs in the s-direction.

$$\Delta V = -\int_{P}^{P+\Delta s} F_s \, ds$$

$$\Delta V(P) = -F_s(P)\Delta s .$$

The changes in the last expression can be compared by taking their ratio:

$$\frac{\Delta V(P)}{\Delta s} = -F_s(P) .$$

This last expression is the change in **V** per unit distance in the **s**-direction (at the point **P**). This is exactly the definition of a derivative except in this case it is written

$$\frac{\partial V}{\partial s} = -F_s$$

where the **script** Greek delta's are used to indicate that this is not the **only** derivative of **V** that exists at **P**. In situations such as this, the derivative obtained is called a **partial derivative.** (*See section* 1.3.3.) The particular partial derivative indicated here is also referred to as a **directional derivative** because the different partial derivatives that are possible at **P** are associated with the different possible directions of differential displacement from **P**. The above remarks apply to any direction **s**. In particular, the **x**-, **y**- and **z**-components of force are

$$F_x = -\frac{\partial V}{\partial x}, \qquad F_y = -\frac{\partial V}{\partial y}, \qquad F_z = -\frac{\partial V}{\partial z}$$

Implied by the previous discussion is the rule for finding partial derivatives. Consider for example, the first of the three relations above. It is necessary to note only that to find $\frac{\partial V}{\partial x}$ the change in **V** was observed **as we moved in the x-direction.** But when one considers movement in the **x**-direction, **y** and **z** are constant. It follows from this that

> *a partial derivative of a given function is found by taking the derivative with respect to the variable indicated treating all other (independent) variables as if they were constants.*

.....**End Mathematical Glitch**

Attention is focused for the moment on systems having only one mass acted on by conservative forces.

Recall that the condition for a system to be in a state of equilibrium is that the net force acting on any part of the system equal zero. It is a simple matter to translate this into a condition on the potential energy:

$$F_x \;=\; F_y \;=\; F_z \;=\; 0$$

$$\Rightarrow \quad \frac{\partial V}{\partial x} \;=\; \frac{\partial V}{\partial y} \;=\; \frac{\partial V}{\partial z} \;=\; 0$$

a point P in space is an equilibrium point if and only if the partial derivative of V with respect to each of the major directions is zero at P.

Now suppose that **P** is an equilibrium point and recall that the condition for stable equilibrium is that small displacements from **P** in any direction be accompanied by a restoring force, i.e. a force that tends to return the mass to its equilibrium position. Displace the mass from **P** calling the direction of this displacement the positive *s*-direction: then the equation

$$F_s \;=\; -\frac{\partial V}{\partial s}$$

indicates that F_s is negative when the slope of V is positive: then the force points in the negative *s*-direction which is back towards **P**. This condition must hold no matter what direction is chosen for *s*. Hence,

a point P in space is a stable equilibrium point if and only if V has a relative minimum at P.

(If V has a relative minimum at **P**, then a displacement Δs in any direction is accompanied by an increase in V which implies $\frac{\partial V}{\partial s}$ is positive and $F_s \left(= -\frac{\partial V}{\partial s} \right)$ is negative. Hence the force is restorative.)

If there is **any** direction *s* such that the derivative $\frac{\partial V}{\partial s}$ at *P* is negative then the equilibrium is unstable. An extreme case occurs when **all** the directions from *P* are accompanied by a decrease in *V*. This is the case when *V* has a relative maximum at *P*.

Therefore

> *an equilibrium point P is unstable if and only if V has a relative maximum, a saddle point or an inflection point at P.*

Finally, if after displacement, there is still no force acting on the body (i.e., it is still in equilibrium) then the derivatives of *V* are zero not only at *P* but also throughout a region containing *P*. It follows that

> *an equilibrium point P is extensive if and only if P is an internal point of some region over which the potential energy is constant.*

These are the rules for finding and categorizing equilibrium points in terms of the potential energy.

2.3.3 Application

Refer once again to *figure* 2.2 noting that the forces are conservative in each of the three cases. Once again, the equilibrium will be categorized in each of the three cases, but this time from the viewpoint of the potential energy. Note that when a mass is constrained to move on a surface, the normal force is always perpendicular to the motion: it therefore does no work and hence does not enter into the energy equation. The potential energy is therefore that of the gravitational field, $V = mgz$, where z is the height above some reference level.

Since V is proportional to the height above the table top, say, it has zero slope where the (plane tangent to the) constraining surface is horizontal. These are the equilibrium points as determined before. In the first case, z and therefore V has a relative minimum so the equilibrium is stable; in the second it has a relative maximum so the equilibrium is unstable; in the third, it is constant so the equilibrium is extensive.

The three cases considered here were chosen for their simplicity so attention could be focused on the principles involved and the methods used rather than on the details of the problem. In a complex system however, the energy relations usually represent a considerable reduction in the work and tedium necessary to analyze the problem. This results from the fact that the potential energy frees you from having to consider the details of the forces involved.

We have conjured up an elephant to crack a peanut here, but not recklessly: in the long run it is more important to become acquainted with the elephant than it is to get the peanut cracked.

2.3.4 Another Static Ride on the See-Saw

As a further example, the moment of mass relation for balancing the see-saw will be derived again, this time using the potential energy instead of the work.

The system is conservative since the only force that does work is the local gravity force. (The unknown forces at the fulcrum are stationary and since they do not act over a distance, they do no work.) We can therefore use the results of the previous section.

The initial potential energy is (rf. to *figure* **2.1**)

$$V_1 \quad = \quad MgH + mgh$$

Using **H** as the independent variable and noting that

$$\Delta h \quad = \quad -k\Delta H$$

where the sign accounts for the fact that **h** decreases when **H** increases, the final potential energy is

$$V_2 \quad = \quad Mg(H + \Delta H) + mg(h - k\Delta H)$$

The change in potential energy is

$$\Delta V \quad = \quad V_2 - V_1 \quad = \quad (M - km)g \, \Delta H \, .$$

The condition for equilibrium is

$$\Delta V \quad = \quad 0$$

$$\Rightarrow \quad M - km \quad = \quad 0$$

$$\Rightarrow \quad M - \frac{r0}{R} \, m \quad = \quad 0$$

$$\Rightarrow \quad MR \quad = \quad mr$$

This is the same result as was obtained before (refer to *section* **2.2.1**). (Incidentally, this result indicates that **V** is constant for any angular displacement: hence, the equilibrium is extensive.)

2.4   THE CENTER OF MASS

2.4.1 Definition of the Center of Mass

It can be expected that the balance point considered in the previous section be an important parameter of any system under investigation. This is in fact the case and therefore we are motivated to make the...

> *definition: the center of mass of a system is that point around which the sum of the mass moments is zero.*

One would eventually define the center of mass after working with a number of systems simply because he would find it and associated expressions continually turning up in his equations. It will become evident as we progress that any complex gyrations of a body can be decomposed into two component motions: the motion of the center of mass and motion around the center of mass. Do not think that this is just a mathematical contrivance (such a decomposition could always be forced on a system mathematically); the rather 0impressive fact is that the dynamics of motion as determined by Newton's equations decompose themselves naturally into these two separate and fairly independent problems and allow for the description of the real motion as their (vector) sum.

2.4.2   Finding the Center of Mass

The following is a derivation of the formula for the coödinates of the center of mass (abbreviated *CM*) of an arbitrary collection of point masses. Let

$$m_1, m_2, m_3, \dots \vec{r}_1, \vec{r}_2, \vec{r}_3, \dots \text{ and } r'_1, r'_2, r'_3, \dots$$

be respectively, an arbitrary collection of point masses, their position vectors referred to the origin and their p0osition vectors referred to the *CM*. Also let *M* be the mass of the entire collection and $\overrightarrow{R}$, the vector from the origin to the *CM*. (See *figure* 2.3.) The following is the derivation of the equation for finding $\overrightarrow{R}$.

The position vector of a typical mass, say the ith, is

$$\vec{r}_i \;=\; \vec{R} \;+\; \vec{r}_i'$$

If each of these vectors is multiplied by its corresponding mass and the resulting equations are added together, the result is

$$\sum_i m_i \vec{r}_i \;=\; \vec{R} \sum_i m_i \;+\; \sum_i m_i \vec{r}_i'$$

Consider each of the two terms on the right separately. In the first, is the sum of the masses, which is the total mass **M** of the system; in the second, is the sum of the moments of mass around the point defined by $\vec{R}$. Recall that the CM is defined as that point around which the mass moments add to zero. Hence, when $\vec{R}$ is the vector from the origin to the CM, the last sum is zero. The position vector $\vec{R}$ of the CM is therefore given by

$$\vec{R} \;=\; \frac{1}{M} \sum_i m_i \vec{r}_i$$

Th0is equation expresses the vector from the origin to the **CM** in terms of the total mass and the moments of mass around the origin. The same point in space is determined no matter where you place the origin or how you orient the axes.

The method for finding the **CM** of a continuous body is the same except that the point masses are replaced by differential chunks of the body and the sum over the point masses is replaced by an analogous integral.

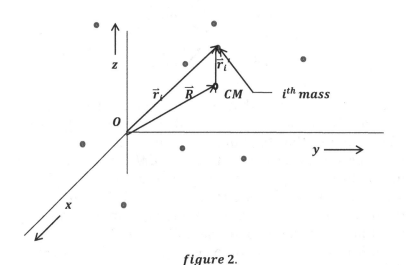

figure 2.

CM of an Arbitrary Conglomerate of Point Masses

2.4.3   The Velocity and Acceleration of the *CM*

The velocity V and the acceleration A of the *CM* are found by taking the time-derivatives of its position vector:

$$\vec{V} \;=\; \frac{d\vec{R}}{dt} \;=\; \frac{1}{M}\sum_i m_i \vec{v}_i$$

$$\vec{A} \;=\; \frac{d\vec{V}}{dt} \;=\; \frac{d^2\vec{R}}{dt^2} \;=\; \frac{1}{M}\sum_i m_i \vec{a}_i$$

These equations can be written in a slightly different form:

$$M\vec{R} \;=\; \sum_i m_i \vec{r}_i$$

$$M\vec{V} \;=\; \sum_i m_i \vec{v}_i$$

$$M\vec{A} \;=\; \sum_i m_i \vec{a}_i$$

Imagine for a moment that you have a point mass **M** at the **CM** moving with velocity **V** and acceleration **A**. These equations say that the sum of the mass moments is the same as the mass moment of the **CM**; the sum of the momenta is the same as the momentum of the **CM**; the sum of the **ma** products is the same as **MA**. This suggests a possible equivalence between the system of arbitrary point masses and the point mass **M** at the **CM**. In the next section, this will be shown to be the case and will in fact be the basis for the extension of Newton's laws from point masses to completely arbitrary systems.

Incidentally, the notation u0sed above will be the same for all the point mass conglomerates that follow. Unless otherwise stated these conglomerates will be completely general.

2.5 THE EXTENSION OF NEWTON'S LAWS

2.5.1 Preliminary Remarks

The statements made at the close of the previous section already suggest the equivalence between a point mass and an arbitrary conglomerate of point masses. This equivalence will now be exploited by considering the most general case.

To be convinced that the following analysis really extends Newton's laws to arbitrary systems, it is necessary to understand what systems can be formed by the point mass conglomerates being considered. Note that:

1. *the number of points is arbitrary;*
2. *each point can be acted upon by any arbitrary external force;*
3. *any two points internal to the system can exert arbitrary forces on each other as long as they are equal and opposite.*

Note further that internal intermolecular forces, electro-static or electro-dynamic forces, gravitational forces, contact forces, etc., taken separately or in combination can hold any conglomerate of point masses together as a solid, liquid, g0as, glue, plasma, galaxy or what-have-you, so it is difficult to imagine anything not included in the generality bragged about here. In any case, whatever can be included will be shown to have its motions governed by Newton's dynamics law and his law of action-reaction. In particular, any collection of solid bodies can be so formed and this minimal appreciation of the generality of which we speak is sufficient to cover the rest of this book.

A restricted version of the results of the analysis will show that the *CM* acts as if:

1. *the entire mass of the system were concentrated at the CM;*
2. *the net external force acts on the CM;*
3. *the motion of the CM obeys Newton's Laws.*

In short, the *CM* moves like a point whose mass is that of the entire system. It is thi0000s fact that germinates the natural decomposition of arbitrary motions into the vector sum of the motion of the *CM* and motion around the *CM*.

In the proof that follows, the notation is the same as in the previous section. Additionally, let $\vec{F}_i$ be the external force acting on the ith mass, and $\vec{p}_i$, its momentum. Referring to the internal forces, let $\vec{G}_{ij}$ represent the force exerted on the ith mass by the jth mass. Finally, let $\vec{F}$ be the net external force and $\vec{P}$, the total momentum.

2.5.2   The Extension Proof

Consider the ith mass and all the forces acting on it. Newton's dynamics equation for this mass is

$$\vec{F} = \frac{d\vec{p}}{dt}$$

$$\vec{F}_i + \sum_{\substack{j \\ j \neq i}} \vec{G}_{ij} = \frac{d\vec{p}_i}{dt}$$

Add all these equations together:

$$\sum_i \vec{F}_i + \sum_i \sum_{\substack{j \\ j \neq i}} \vec{G}_{ij} = \frac{d}{dt} \sum_i \vec{p}_i$$

The first term is $\vec{F}$, the sum of the external forces. The second term refers to the internal forces. But if body j exerts a force on body i, then body i exerts and equal and opposite force on body j, (law of action-reaction). Therefore

$$\vec{G}_{ij} = -\vec{G}_{ji} \qquad if \ i \neq j$$

$$G_{ii} = 0 \qquad for \ all \ i$$

The last condition precludes the possibility that a mass can exert a force on itself (or equivalently, that I can pick myself up by my bootstraps). The result of all this is that the $\vec{G}'s$ cancel themselves out in pairs and the second term is zero. The order of summation and differentiation were interchanged in the term on the right: this is legitimate because the sum of the derivatives is the derivative of the sum. The term on the right therefore is the time-derivative of the total momentum. The equation above then becomes

$$\vec{F} \;=\; \frac{d\vec{P}}{dt} \;=\; M\frac{d\vec{V}}{dt} \;=\; M\vec{A}$$

where $\vec{F}$ is the net external force, $\vec{P}$ is the total momentum, M is the total mass, V is the velocity of the CM and A is the acceleration of the CM. Thus the CM of any arbitrary conglomerate of point masses satisfies Newton's laws,(Included is any arbitrary solid which is a conglomerate of its differential pieces.)

This establishes the result. Hence

> *the assumption that Newton's laws of motion govern the motion of point masses implies that they also govern the motion of any arbitrary mass.*

Keep in mind that this includes any extended body or collection of extended bodies solid, liquid, gas or combinations thereof, etc: therefore, for situations outside the realms of relativity and quantum mechanics,

> *Newtonian mechanics is a complete world view of the motions of physical objects.*

Note that in the proof given above, the internal electro-dynamic forces still cancel each other: hence, the extension applies even in the presence of these forces. Recall that the stipulation was that when they are present, Newton's action-reaction law be relaxed a little so that it does not include the 'along-the-line-of-centers' part, i.e. the forces two bodies exert on one another are equal and opposite but not necessarily along their line of centers. This stipulation has no effect on the foregoing analysis.

CHAPTER III

Applications to Linear Motion

3.1 SOME GENERALIZATIONS OF LINEAR MOTION

3.1.1 Preliminary Remarks

The results of the preceding chapters will now be used to describe some of the aspects of linear motion. Although the results will be quite general, attention will be confined for the most part to the motion of solid bodies and to situations where they are not deformed. (An exception to this is the analysis of inelastic collisions which involve the deformation of at least one of the bodies.) Liquids and gases will be considered only peripherally as their analyses contain the mathematical difficulties of the changes in shape and volume that accompany their motion.[7]

The issues of shape and volume changes require the use of vector-tensor calculus for any compact description. It is emphasized here however that even with all these intricacies, each differential volume of fluid moves, changes shape, compresses and expands according to the dictates of $\vec{F} = m\vec{a}$. Indeed the gas law itself is derivable from the dynamics equation: the derivation belongs to a subject called statistical mechanics where it is seen that

---

[7] Each differential volume of either a liquid or a gas changes the shape of its boundary as it moves. Liquids are only slightly compressible but this can often be neglected; gasses, on the other hand are highly compressible, each differential region adjusting its volume (and temperature) to accommodate the current pressure in its vicinity. For an ideal gas, the pressure, volume and temperature are related by the gas law $PV = nRT$ where the pressure-volume product is exhibited as proportional to the temperature.

the gas law results when the statistics of the motions of the individual molecules of the gas are considered, each moving according to the Newtonian prescriptions.

The restriction of our attention to the motion of solid bodies will enable us to glean some of the immediate consequences of the previous chapters. For example, it was determined that if one were to consider the mass *M* of an entire system as concentrated at its *CM*, then the *CM* would move as if it were a point mass *M* acted upon by the (vector) sum of all the external forces. The motion of the *CM* is therefore independent of whatever internal forces are in play.

Understand that the tasks of determining the position and orientation of a solid body can be difficult. The remark of the preceding paragraph however constitutes a major step in the breakdown of the general problem into simpler ones. For example, consider a system consisting of many bodies with arbitrary interactive forces between them: if the system is not acted upon by any external forces, its *CM* moves according to the law of inertia, i.e., in a straight line with constant speed. The solar system is an example. (The large distances separating it from external influences on its motion render their effects negligible.) The *CM* of the solar system therefore moves as described above. If the solar system were to explode into cosmic dust tomorrow, the *CM* of the resulting particles would continue to wend its way in the same fashion completely unperturbed.

Note that by considering the solar system *as a unit*, we determine its motion *as a unit*. Now suppose we want to determine the details of the earth-moon system (neglecting the other planets whose effects are small perturbations compared to the effect of the sun). We define these two bodies as the system but when we do so, the gravitational attraction of the sun on this system (an *internal* force when the solar system was considered as a unit), is *external* to the earth-moon system. Then the orbit of the *CM* of the earth-moon system around the sum can be determined. Once this is found, we can go further and consider the earth alone, defining it to be the only mass in the system. Then the moon's gravitational force (previously *internal*) becomes an *external* force and the *CM* of the earth moves around the *CM* of the earth-moon system as if it were a point, whose mass equaled that of the earth, acted upon by the moon's gravitational pull.

Thus the question "How does Peoria, Illinois move with respect to the distant galaxies?" is partially answered as a sum of three separate motions: the motion of the *CM* of the solar system, plus (that's a vector-type 'plus') the motion of the *CM* of the earth-moon system around the *CM* of the solar system, plus the motion of the earth's *CM* around the *CM* of the earth-moon system. (We could complete the problem and consider the motion of the *CM* of Peoria, Illinois around the earth's *CM* but the point is already made.)

The fact that the CM of a system moves as if it were a point mass satisfying Newton's laws affords a doorway into the solution of complex problems. Note that at each stage of refining the problem, the choice of which constituents are to become internal to the syste0m is arbitrary and can be made in any fashion convenient to the problem. Once made however, the choice determines which forces are internal and which are external. The subsequent analysis refers to the motion of the CM of

the system chosen under the influence of
the forces external to it.

3.1.2 The Conservation of Momentum

Recall from the last chapter that

$$\vec{F} \;=\; \frac{d\vec{P}}{dt} \;=\; \frac{d}{dt}\sum_i m_i \vec{v}_i\,,$$

where $\vec{F}$ is the net external force and $\vec{P}$ is the momentum
of the **CM** found to be equal to sum of the momenta of the
individual point masses. Applying Newton's law to this
equation, the absence of a net external force implies that $\vec{P}$,
the momentum of the **CM** is a constant (vector). The same
result was obtained for a single point mass in the absence
of a net external force. We have

> *The Conservation of Linear Momentum:*
> *the CM of any system not subject to a net*
> *external force follows the law of inertia.*
> *This is equivalent to the statement that the*
> *(vector) sum of the momenta of the*
> *individual parts of the system is a constant*
> *(vector). In mathematical symbols,*
>
> $$\vec{P} \;=\; \sum_l m_i \vec{v}_i \;=\; \vec{K}$$
>
> *where $\vec{K}$ is a constant (vector).*

(Thus the sum of the momenta of all the objects in the
solar system adds up to a constant vector. The generality
of these principles makes sweeping statements like this
possible.

The conservation of momentum constitutes the basis for the discussion of collision phenomena where neither of two colliding bodies is subject to a net external force. In connection with the analysis of collisions, it is often expedient to express all quantities in the *CM* system. This is a system whose origin is at the *CM* for the duration of the motion.

The freedom to transfer from one coördinate system to another entails a discussion of the equivalence of inertial frames of reference and the development of the formulas which relate quantities like momentum and kinetic energy as measured in one system to the same quantities as measured in another.

Accompanying this discussion is the development of some more vector algebra: two types of vector products are presented:

1. **the dot product (aka the inner product, aka the scaler product) which combines two vectors to form a scaler;**
2. **the cross product (aka the vector product) which combines two vectors to produce a third.**

The collision problems already analyzed are then reconsidered, but this time the analysis is carried out in the *CM* system. A comparison of the two analyses will illustrate the simplifications brought about by the use of the *CM* system.

3.1.3   Motion of a Body in a Constant Force Field

The next topic presented is the motion of bodies in a constant force field. There are two types of investigations:

the first includes trajectory problems near the surface of the earth (in the local gravitational field) where a body is in free flight after it is given an initial velocity; the second includes problems involving masses, inclined planes, strings, pulleys and the like.

In this latter ilk of systems, the motion is constrained by the geometry of the system and these constraints give rise to unknown forces. The problems however, are usually solvable in spite of this by virtue of the fact that these forces do no work and hence do not affect the kinetic energy. This is one of the simplifications that results from the consideration of frictionless surfaces. For example, a block sliding down an inclined plane is constrained to move along the surface of the plane; this is a geometric constraint which normally gives rise to a force unknown in both magnitude and direction where the block and the plane are in contact. The assumption that the surfaces are frictionless however means that the unknown force cannot have a component parallel to the constraining surface which is the direction of motion, i.e.

> *frictionless surfaces can sustain only a normal component of contact force.*

Well, isn't that nice! The normal component is always perpendicular to the direction of motion and therefore does no work. It's almost as if it were planned that way. Since the only force that does work in these systems is the local (constant) gravity force, the total energy is conserved. It follows that when there is no force in the direction of motion, the constraint affects only the direction of the velocity and not its magnitude.

Recall that the work done by a force is the product of the component of force in the direction of motion and the distance through which it moves. In the above case there is no component of force in the direction of motion. In contradistinction, there is an important case where there is no motion in the direction of force. This sounds like a facetious logical equivalent to the last statement but it is not quite that; the statement affords an understanding not imparted by the previous one.

The reference here is to a *rolling friction force.* If the surfaces are truly frictionless, then a sphere or a cylinder would slide rather than roll down an inclined plane. (The object in question might exhibit some initial circular motion but this would not be related to its contact with the plane.)

The situation might be described this way. Imagine a sphere rolling down an inclined plane. The friction force is parallel to the plane and is applied at the point of contact. But picture the differential piece of the sphere where it touches the plane: this is the only part of the sphere subject to the friction force and this piece merely touches the plane, stays in *stationary* contact for an instant and then lifts off. As long as there is no slippage, the actual mass point subject to the friction force never moves in the direction of the force. Therefore the rolling friction force does no work.

What then is the effect of the rolling friction force? It certainly cannot result in the same motion as the frictionless case. We already know one difference between the two: the object slides over a frictionless surface and .rolls over a non-frictionless one. The answer lies in this difference. In each case, energy is conserved so the kinetic energy attained by the sphere after descending a vertical height *h* is *mgh* in both cases. In the case of rolling, part of this energy is rotational (motion around the sphere's center) and part is translational (motion down the inclined plane). The friction force is precisely that necessary to apportion the available energy *(mgh)* to these two types of motion in such a way that the linear and rotational motions match at the point of contact and there is no slippage.

3.1.4   Linear Restoring Force

The last topic to be considered in this chapter is motion in one dimension under the action of a linear restoring force. This is essentially the study of oscillatory motion and provides the groundwork for the investigation of all wave phenomena whether it be sound waves, light waves, water waves or the matter waves of quantum mechanics.

The oscillatory motion results when a mass displaced from its equilibrium experiences a force which tends to return it. This is referred to as a ***restoring force.*** When this restoring force is proportional to the displacement from equilibrium, it is called a ***linear restoring force*** and in this case the equations of motion take on a particularly simple form and can be solved. Springs exhibit this kind of a restoring force: if a spring is elongated or compressed it delivers a force equal to *(–kx)* where the minus sign indicates that the force is restoring (when *x > 0,* it points in the negative direction and visa versa), *k* is a constant related to the stiffness of the spring and *x* is the displacement from the equilibrium position.

The one-dimensional motion of a mass attached to a spring will be considered and it will be found that the solution of $\vec{F} = m\vec{a}$ for this system is a sinusoidal oscillation. The parameters of this kind of motion (amplitude, frequency, period, angular velocity, etc.) pervade all wave motion and hence deserve some attention.

Note that once again we are dealing with a conservative force field (refer to *section* **1. 3. 4. 2**):

$$\vec{F} = -kx$$
$$\Rightarrow V = k \int x \, dx$$
$$\Rightarrow V = \tfrac{1}{2}k x^2 .$$

A short trek into mathematics ----- the extension from simple harmonic motion (sinusoidal oscillations) to any arbitrary periodic motion was the work of a physicist named Fourier who initiated the branch of mathematics known as *Fourier analysis.* He needed the results for his work with the heat equation. His problem was to determine the temperature distribution throughout a region when the temperature distribution at every point on the bounda.ry of the region is known. (This is called a boundary value problem. Without knowing the details, it is difficult to understand what this has to do with oscillatory functions --- but that's another story.)

His fundamental result was that any periodic function (as long as you don't get weird and pick functions that go to infinity or are discontinuous at every point, etc.) can be written as an infinite sum of sines and cosines all of whose terms have frequencies which are integral multiples of that of the given function. (Some of the terms may have zero coefficients, but the point is that sinusoids with *only* these frequencies appear in the sum.)

The analysis produces the expression

$$F(x) = \frac{4}{\pi}\left[sin(x) + \frac{1}{3}sin(3x) + \frac{1}{5}sin(5x) + \cdots\right].$$

for the square pulse shown in *figure* **3. 1.**

Chapter III: Applications to Linear Motion

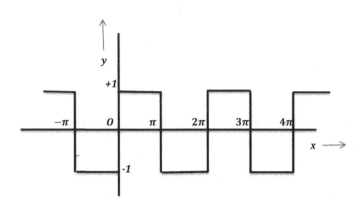

figure 3.1

Fourier Analysis

EXERCISE: On a piece of graph paper, make a fairly accurate plot of the first term; then using the same coödinate axes, plot the second and third terms. Finally, verify the validity of the above equation by taking increments along the x-axis and at each of these values of x add the three heights, graphically forming their sum. After only three terms, the result should already approximate the square pulse of the figure. Note that after 2π both the square pulse and the sine waves repeat thus forming a periodic function that goes on forever.

The chapter concludes with an analysis of the phenomenon of resonance. Resonance occurs when an oscillating system experiences an external excitation whose frequency is at or near the natural frequency of the system. The system responds wildly and in the absence of sufficient damping, it literally explodes. The equations which describe this phenomenon are derived from the dynamics equation followed by a detailed analysis of the results.

At this point the three most elementary types of force will have been introduced: the zero field, the constant field and the linear restoring field.

3.2 SYSTEMS WITH NO NET EXTERNAL FORCE

3.2.1 The Equations for the Collision of Two Bodies

Consider two bodies on a horizontal frictionless surface where the normal forces exactly balance the weights. If it is noted that there are no forces acting in the horizontal direction, and that those in the vertical direction add to zero for each of the bodies, it can be concluded that there is no net force on the system. The table represents a constraint which serves only to limit the movement of the bodies to the two-dimensions of its surface.

The initial situation is depicted in *figure* $3.2.a$ below. A mass m_1, is travelling along the negative x-axis toward the origin with a velocity v_1. A second mass m_2 is at rest at the origin. They collide at time $t = 0$. Nothing is known about the forces they exert on each other at the instant of collision. Whatever can be said about the system after collision follows from the conservation of momentum. (Momentum is conserved because there is no external force on the system.)

Let v_1' making an angle θ with the positive x-axis and v_2' making an angle φ with the positive x-axis (as shown in the figure) be the velocities after collision. Let $\vec{P}$ and $\vec{P}'$ be the initial and final momenta of the system. When the conservation of momentum is applied to the system, we get two equations, one for each direction:

$$\vec{P} = \vec{P'}$$

(x-dir.): $\quad m_1 v_1 = m_1 v_1' \cos(\theta) + m_2 v_2' \cos(\varphi)$

(y-dir): $\quad 0 = m_1 v_1' \sin(\theta) - m_2 v_2' \sin(\varphi)$

The two angles and the two final speeds are unknown. We therefore have **2** equations and **4** unknowns. There is not enough information to solve the equations for the final state. Usually when this type of experiment is carried out, the speed and deflection angle of one of the masses is observed after collision. Thus two of the unknowns are determined by direct measurement: then what remains is a set of 2 equations in 2 unknowns and this is sufficient to determine the speed and deflection angle of the second mass.

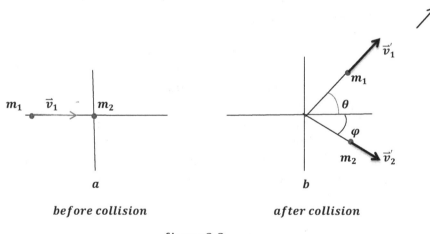

a   b

before collision   *after collision*

figure 3.2

Two Body Collision

Two extreme cases will be considered: perfectly *inelastic* collisions and perfectly *elastic* collisions. In the following section, there is a discussion of the former (these are collisions in which the bodies adhere to each other upon impact): in this case both masses have the same deflection angle and the same final velocity. Subsequently, there is a discussion of perfectly elastic collisions (these are collisions in which the contact forces at the instant of impact are conservative): in this case energy is conserved. (That's what *'perfectly elastic'* means.)

3.2.2 Perfectly Inelastic Collisions

As mentioned above, a perfectly inelastic collision is one in which the two bodies adhere to each other upon impact. (We are considering the case when the dimensions of the bodies are small. This will be made more general later.)

We get two equations from the conservation of momentum and two additional ones from the adherence of the bodies:

Conservation of momentum

x-direction:
$$m_1 v_1 = m_1 v_1' \cos(\theta) + m_2 v_2' \cos(\varphi) \quad (1)$$
y-direction:
$$0 = m_1 v_1' \sin(\theta) - m_2 v_2' \sin(\varphi) \quad (2)$$
Adherence of the bodies
$$v_1' = v_2' \quad (3)$$
$$\varphi = -\theta \quad (4)$$

The first two are the conservation of momentum equations (which we had earlier); equations (3) and (4) reflect that the bodies adhere to each other after collision. The minus sign in equation (4) results from the fact that φ and θ were defined to be in opposite directions (see *figure 3.2.b*). This fact was already written into equation (2) and consistency requires that it not be redefined when we get to equation (4). Using the last two conditions, the first two become

$$m_1 v_1 \quad = \quad [m_1 + m_2]v_2' co\, s(\theta)$$
$$0 \quad = \quad [m_1 + m_2]v_2' si\, n(\theta)$$

In the last equation, at least one of the three factors on the right must be zero. Since each of the first two is non-zero, we have

$$sin(\theta) \quad = \quad 0$$
$$\Rightarrow \quad \theta \quad = \quad 0$$
$$\Rightarrow \quad \varphi \quad = \quad 0$$

Hence, $cos(\theta) \ = \ 1$ and equation (1) becomes

$$v_2' \quad = \quad \frac{m_1 v_1}{m_1 + m_2}$$

The expression on the right is the same as that obtained earlier for the velocity of the *CM* along the *x*-axis. This is exactly what we would expect since both bodies after collision are and remain at the *CM* and its velocity after collision must be the same as it was before collision.

The situation is depicted in *figure 3.3*. Note that since we started with four equations in four unknowns, we were able to obtain a complete solution.

Consider the more general situation when the bodies are not small. *Figure 3.4* shows the collision of two extended masses. Their centers of mass are shown with **CM_1** moving directly toward **CM_2**. As long as the two centers of mass lie on the line that represents the path of motion, the motion is essentially the same as that found above: the two masses adhere upon impact and continue straight along the x-axis.

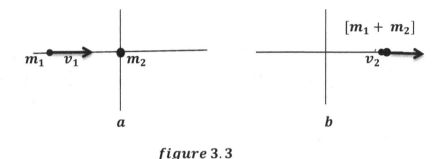

figure 3.3

Inelastic Collision between Two Bodies

A more general situation is depicted in *figure* 3.5: *CM_1* is not lined up prior to impact so that when the masses collide, the second mass experiences not only a linear impulse in the *x*-direction but also an angular impulse around its center of mass. The same is true for the reaction impulse on *CM_1*. The net result is that after impact the two bodies adhere to one another and exhibit both translational and rotational motion. Their net motion is the vector sum of these.

It might be suspected that energy is lost in a perfectly inelastic collision because the adherence of the two masses upon impact involves some deformation and heat dissipation. This is in fact the case, but the details of two body collisions are more easily comprehended when viewed from a special frame of reference, namely, the CM system. This is presented later in section 3.2.8.1. The question of energy loss will be discussed there.

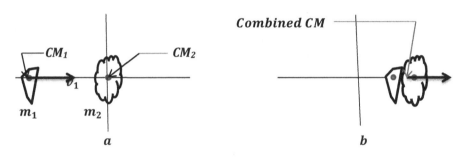

figure 3.4

Inelastic Collision of Two Extended Bodies, Aligned

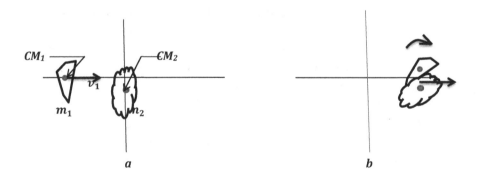

figure 3.5

Inelastic Collision of Two Extended Bodies, Not Aligned

Where did the energy go? To answer this, you must look at the details of the impact-adherence process. You could be dealing with sticky surfaces, deformations of the bodies, chemical bonding, the capture of an electron by an atom, or any number of such things. These processes usually dissipate heat and/or light.

It is important to realize however, that in classical physics, energy is never 'lost'. When we say that it is, we mean that it has been transformed into another form of energy which renders it no longer available to the motion of the system. The heat dissipation at the point of contact (which affects the adherence) manifests itself as an increase in the oscillatory motion of the local molecules which in turn, exert local forces on their neighbors. These are the internal G_{ij} forces of *section* **2.5.2** which as shown there can no longer affect the motion of the system.

3.2.3 Perfectly Elastic Collisions

It will be shown later that a perfectly inelastic collision results in the maximum possible loss of kinetic energy. A perfectly elastic collision is the antithesis of this: the contact forces at impact are conservative and hence there is no energy loss. Anyone who has played marbles or billiards has a sense of elasticity. Basically, the elasticity of a material refers to its ability to retain its shape and to reclaim it if slightly deformed. Any material with an inability to restore its shape after impact suffers a permanent deformation which is always accompanied by the dissipation of heat and hence a loss of (mechanical) energy. In contrast, highly elastic materials upon impact depress slightly and then restore their original shape. In other words they act like springs which compress at impact changing the energy from kinetic to potential and then return to their previous shape thus changing the energy back to kinetic. This is the process of bouncing off each other, Materials like high-grade spring steel come close to perfect elasticity.

The equations for an elastic collision are

conservation of momentum

x − direction

$$m_1 v_1 \;=\; m_1 v_1' \cos(\theta) + m_2 v_2' \cos(\varphi) \qquad (1)$$

y − direction

$$0 \;=\; m_1 v_1' \sin(\theta) - m_2 v_2' \sin(\varphi) BI \qquad (2)$$

Conservation of energy

$$\tfrac{1}{2} m_1 v_1^2 \;=\; \tfrac{1}{2} m_1 v_1'^2 + \tfrac{1}{2} m_2 v'^2_2 \qquad (3)$$

where the first two equations are the momentum equations (which we had before) and the third is the conservation of energy equation.

We now have three equations in 4 unknowns so they are not completely solvable. There is an added difficulty: the unknown speeds after collision appear to the first power in the first two equations and as squares in the third. Hence, we do not have a set of *linear* equations. When the equations are not linear, there is no organized way to solve the set or even to extract useful information from it. Basically, you fiddle around trying all possibilities until you eek out something useful.

Sets of non-linear equations often occur in physics, this particular set because both momentum and energy are conserved: momentum is a linear function of the speed; kinetic energy is a quadratic function of the speed.

As an illustration of the type of calculation involved, the set will be reduced to express the relation between the unknowns in a particular way which will enable some statements about the final state of the system. In particular, it will be shown that when the masses are equal, the angle between their final paths of motion is a right angle. However, the masses will be left arbitrary until the end. This will allow for additional statements about the final state.

The reduction of the three equations can be carried out as follows.

Square each of the first two equations:

$$m_1^2 v_1^2 = m_1^2 v_1'^2 \cos^2(\theta) + m_2^2 v_2'^2 \cos^2(\varphi) +$$
$$2 m_1 m_2 v_1' v_2' \cos(\theta) \cos(\varphi) \quad (4)$$

$$0 = m_1^2 v_1'^2 \sin^2(\theta) + m_2^2 v_2'^2 \sin^2(\varphi) -$$
$$2 m_1 m_2 v_1' v_2' \sin(\theta) \sin(\varphi) \quad (5)$$

Add them together and use the two trig identities:

1. $\cos^2(x) + \sin^2(x) = 1$
2. $\cos(\theta + \varphi) = \cos(\theta) \cos(\varphi) - \sin(\theta) \sin(\varphi)$

The result is

$$m_1^2 v_1^2 = m_1^2 v_1'^2 + m_2^2 v_2'^2 + 2 m_1 m_2 v_1' v_2' \cos(\theta + \varphi)$$

The final steps in this process are: multiply **eq (3)** by $2m_1$:

$$m_1^2 v_1^2 = m_1^2 v_1'^2 + m_1 m_2 v_2'^2 \quad (7)$$

and subtract **eq (7)** from **eq (6)**. An $m_2 v_2'$ cancels out and the result is

$$0 = [m_2 - m_1] + m_1 v_1' cos(\theta + \varphi) \qquad (8)$$

Note that in eq **(8)** none of the unknowns were eliminated; they have simply been related in another way. Nevertheless it is still possible to extract some information. Let $m_1 = m_2$, then the first term is zero and

$$m_1 v_1' \, cos(\theta + \varphi) = 0 \qquad (9)$$

It follows that either $v_1' = 0$ or $cos(\theta + \varphi) = 0$. The first of these conditions can be satisfied only if the bodies hit head on in which case m_2 simply acquires the velocity of m_1 and continues along the x-axis; the second condition must hold generally and implies that

$$\theta + \varphi = \frac{\pi}{2}$$

This is the result that was indicated before we began: that the angle between the final directions of motion is **90°**.

Additional information can be obtained if **eq (8)** is put in the form

$$\frac{v_2'}{v_1'} = \frac{cos(\theta + \varphi)}{1 - m_2/m_1}$$

The $v's$ are speeds, so the left side is always positive. In order that the right side be positive, it is necessary that:

$$\theta + \varphi > \frac{\pi}{2} \quad \text{if} \quad m_2 > m_1;$$

$$\theta + \varphi < \frac{\pi}{2} \quad \text{if} \quad m_2 < m_1.$$

It is not clear how one proceeds to obtain these results. The above analysis took a number of tries and was aided by a modest amount of algebraic/trigonometric foresight acquired from the experience of having done this sort of thing before. This is the 'fiddling around' process when a set of equations is not linear.

Exercise: The first result above could have been obtained more easily by assuming the masses equal at the start. Make this assumption and simplify equations **(1)**, **(2)** and **(3)**, appropriately. Then use the analysis above as a guide and go through the steps as they appear with this assumption to obtain the first result, that the angle between the final paths is a right angle.

3.2.4   The Equivalence of Inertial Frames of Reference

Inertial frames of reference are coördinate systems whose acceleration is zero. It will be shown in this section that all inertial frames of reference are equivalent in the sense that Newton's laws are valid in all of them. This will allow the freedom to view problems from a number of different coödinate systems. Consider for a moment, a case where the external force on a system is zero. Then the *CM* moves with constant velocity and therefore any coördinate system at rest with respect to the *CM* is an inertial system. It often happens that problems are considerably simpler when viewed from a *CM* - system. (This is a system whose origin is at the *CM* throughout the motion.)

A *CM* coödinate system can still be of use even when it is accelerating. In fact, it is a central tool in breaking down complex motions into two simpler ones: the motion of the *CM* and motion around the *CM*. (See *section* 3.1.1.) In *section* 3.2.8.2, the problem discussed in the previous section will be translated into the *CM* system. It will be seen that the questions asked about the motions of the two masses can be solved by inspection in this system and that a translation back into the given coördinate system where the second mass was initially at rest will produce the results in the form obtained above.

Consider two inertial frames of reference (two separate coördinate systems neither of which is accelerating, but may have a constant velocity with respect to each other). Suppose that you are at rest with respect to one of them. (We will refer to this one as the stationary system even though 'stationary' has no meaning in this context.) Suppose further that their corresponding axes are parallel and the second moves with a constant velocity $\vec{V}$ with respect to the first. There is no loss of generality if it is assumed that their origins coincide at time $t = 0$. Refer to *figure* 3.6.

In what follows, the unprimed quantities refer to the stationary system and the primed quantities refer to the moving system.

At some arbitrary time t, the moving system is displaced from the stationary one by a distance $\vec{V}t$. It is obvious from the diagram that the position vector $\vec{r}$ of the point P as seen from the stationary system is different from the position vector $\vec{r}'$ as seen from the moving system: the relation between them is

$$\vec{r} \;\; = \;\; \vec{V}t \;\; + \;\; \vec{r}'$$

This is known as the Galilean transformation.[8]

---

[8] This is the point where the special theory of relativity departs from Newtonian mechanics. The hidden assumption in the simple relation cited is that observers in each inertial frame agree on their measurements of time duration and length.

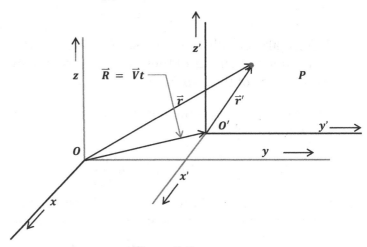

figure 3.6

Inertial Frames of Reference; the Galilean Transformation

The relation between the velocities as measured in the two systems is obtained by taking the time-derivative of the above equation:

$$\frac{d\vec{r}}{dt} = \frac{d}{dt}(\vec{V}t + \vec{r}')$$

$$\vec{v} = \vec{V} + \vec{v}'$$

Similarly, the relation between the accelerations is obtained by taking the next time-derivative:

$$\vec{a} = \frac{d\vec{v}}{dt} = \frac{d}{dt}(\vec{V} + \vec{v}')$$

Special relativity demonstrates that this is only approximately true with the discrepancy0 increasing as $|\vec{V}| \rightarrow c$, the speed of light.

$$\vec{a} = \frac{dV}{dt}$$

$$\vec{a} = \vec{a}'$$

Things are simple enough so far. The next task is to determine the effects of these relations on the dynamics as seen in each of the two frames.

Let **m** be the mass of the body. In the stationary system, we know that

$$\vec{F} = m\frac{d\vec{v}}{dt}$$

If the velocity relation above is used, the dynamics equation can be expressed in terms of the moving system:

$$\vec{F} = m\frac{d\vec{V}}{dt} + m\frac{d\vec{v}'}{dt}$$

$$\vec{F} = 0 + m\frac{d\vec{v}'}{dt} = m\vec{a}'$$

This last relation indicates that Newton's dynamics equation holds in the moving system. (Note that this is not true if $\vec{V}$ is changing with time, i.e., if the second system is accelerating.) All non-accelerating reference frames are given a special name: they are called ***inertial frames of reference***. The result we have just obtained is

> *Newton's laws are valid in all inertial frames of reference.*

Another way to say this is, "As far as mechanics is concerned, all inertial frames of reference are equivalent".

3.2.5 The Center of Mass System

Some problems are considerably simplified when viewed from the *CM* system. This is the system whose origin is at the *CM* and which moves with it without rotating (i.e., the axes move parallel to themselves). For the most part, these simplifications occur whether the *CM* system is inertial or not. (When it is not, its acceleration alters the dynamics law and the altered version must be used when viewing the problem from this system. There will be more about this in the discussion of *d'Alembert's principle.* (See *section* 5.5.) For now, we are interested only in the case that the *CM* system is inertial.)

It may have been expected that this system would render many analyses simpler. First, it is the dynamic mass center of the system. That is how it was defined; Second, some previous results have already indicated that the system acts as though it were a point mass concentrated at the *CM* and acted upon by the sum of all the external forces.

Attention is drawn to the fact that by the definition given here, there are many *CM* systems. While we have been talking about 'the *CM* system', we should have been saying 'a *CM* system'. (If you have one *CM* system, any other coördinate system whose origin coincides with that of the first is also a *CM* system.) There is more to this subject than meets the eye: e.g. solid objects generally have a natural set of axes which can be used to define a particularly useful *CM* system, but this issue is best approached in a course in mathematical physics.[9] For purposes here, there will be no cause for confusion: the

---

[9] The subject of the natural coördinate system and the natural units of length associated with a particular solid body is touched upon in *sections* 5.4.2.2 through 5.4.2.5.

coördinate axes for any *CM* system we will encounter will be obvious.

The task at hand is to develop the formulas which enable us to express the dynamic quantities of mechanics in both the stationary and *CM* systems and to relate them. It is also necessary to be able to switch from one to the other whenever it is expedient to do so. Since vector relations make this task both simpler and more comprehensible, it is no longer feasible to avoid...

3.2.6   The Algebra of Vectors

3.2.6.1   Outline

The material in this section reiterates some of the material in *section* $0.2.3$. A review of that section might be helpful at this point.

Vectors are used extensively in physics. They are the natural descriptors of objects which exhibit the properties of both magnitude and direction. In the simplest applications, they provide an organized description of the position and orientation of bodies as well as a concise and condensed description of their dynamics. In a subtle way, the economy of expression that vectors (also matrices and tensors) embrace allows for the flow of more and more information into fewer and fewer symbols. In this context it is generally true that vector formulations direct attention at the central issues stripping them of extraneous and non-essential considerations. In this way vectors foster the growth of the more abstract and generalized concepts which govern the topics at hand. For, example, to understand the tensor equation

$$G^{\mu\nu} \;=\; 0,$$

is to have a good basic understanding of general relativity. It could very well be that the economy of expression afforded by vectors is their greatest contribution to the furtherance of physics.

The material that follows is associated with switching coördinate systems and later with the analysis of circular motion. It is more easily understood in terms of vectors. Hence, the inevitable task of learning to manipulate them is upon us.

Pertaining to vector algebra, the sections that follow will address:

1. *representation of vectors in Cartesian coordinates;*
2. *a discussion of vector operations;*
3. *addition and subtraction of vectors;*
4. *dot products;*
5. *cross products;*
6. *the basic calculus of dot and cross products.*

3.2.6.2 Representation of Vectors in Cartesian Coördinates

For the most part the vector equations of physics can be derived and manipulated without a coördinate system, but at some point, because their application requires the particulars of a situation, a measuring device, namely a coördinate system must be introduced. The vectors must then be related to this coördinate system to facilitate the calculations appropriate to the current issue. We now discuss the details of representing vectors in a Cartesian coördinate system.

The most basic requirement is the representation of a vector from the origin to a given point (x, y, z). To this end, three auxiliary unit vectors (vectors of length **1**) are defined, one for each of the three principle directions. For the x-direction, let $\vec{\imath}$ represent the vector from the origin to the point $(1, 0, 0)$; similarly, let $\vec{\jmath}$ and $\vec{k}$ go from the origin to the points $(0, 1, 0)$ and $(0, 0, 1)$ for the y- and z-directions, respectively. Then any given vector $\vec{R}$ can be represented as the (vector) sum of three mutually perpendicular vectors, one in each of the principle directions:

$$\vec{R} \;=\; R_x\vec{\imath} \;+\; R_y\vec{\jmath} \;+\; R_z\vec{k},$$

where the coefficient of each unit vector is the projection of $\vec{R}$ onto the corresponding coördinate axis.

This last assertion pre-empts the following sections a little, but it will become clear that there has been no harm done and all is consistent (once the addition of vectors is defined).

Before continuing, we note that the vector $\vec{R}$ is the diagonal of a parallelepiped with sides R_x, R_y, and R_z: using the Pythagorean Theorem, it can be seen that the magnitude (length) of $\vec{R}$ is

$$|\vec{R}| \;=\; \sqrt{R_x^2 \;+\; R_y^2 \;+\; R_z^2}$$

3.2.6.3 Vector Operations

At the onset, it is important to understand what constitutes a vector operation and what does not. Keep in mind that vectors are used to represent physical quantities in space. In order to quantify these descriptions, a coördinate system is introduced: it serves to locate points, track time variations, represent vectors and their interactions, etc. If one coördinate system is removed and another inserted, none of the physical quantities represented change from what they were before the switch: these quantities are the same as they were although they may not look the same when expressed in the new coördinate system.

The previous statement is the crux of the matter under discussion: in order that an equation or operation be acceptable as a description of a physical situation, all physical quantities represented must remain *essentially invariant* when the coördinate system is changed. The italics refer to the fact that the quantities may look different in the new representation but the new system must, regardless of this apparent difference, represent the same physical objects. For example, if the vector from the origin to the point $(1, 3)$ in the xy-plane is expressed in a new coördinate rotated **90°** counter-clockwise from the old one, the new representation of the same vector is the vector from the origin to the point $(3, -1)$. This is the same vector in space that we started with, so there is no essential change even though it looks different.

EXERCISE: Make a sketch illustrating the above description. Show the vector and the two coördinate systems and convince yourself that the two representations depict the same object.

It is not a simple matter in general to recognize whether or not a given operation is a legitimate vector operation. Essential invariance is the central issue here and it is necessary to ensure that the vector operations we will define are compatible with the invariance principle. Therefore, the following criterion is adopted:

> *if an operation involving vectors is to be accepted as a legitimate vector operation then there must exist at least one definition of the operation which references the vectors only and in no way references any system of coördinates which may be used to represent them.*

A little thought reveals that any operation which meets this criterion can produce only results that are independent of any coördinate system.

The above remarks require a little more explanation. Note first that sometimes an operation results in a scaler and sometimes it results in a vector or some other mathematical object. In the former case, a legitimate vector operation when carried out will always produce the same scaler no matter what coördinate system is used to represent the vectors. An illegitimate operation will not. Recall that scalers are just numbers and therefore are not affected by a change of coördinate systems. In the latter case, the result of involving the component vectors in a legitimate operation will generally produce vectors that look different but are essentially the same as in the exercise above; the result of involving the component vectors in an illegitimate operation will generally produce vectors that look different and are essentially different. Clearly, in the latter case it is difficult to distinguish between legitimate and illegitimate operations and the criterion cited above must be relied upon.

An example of a legitimate vector operation is taking the magnitude of the vector. First, we present a definition which does not reference any coördinate system:

> *given a vector $\vec{V}$, its magnitude is its length, denoted $|\vec{V}|$.*

This definition does not refer to any coördinate system and therefore satisfies the criterion cited above: hence it is a legitimate vector operation.

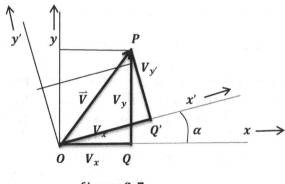

figure 3.7

The Invariance of Vector Magnitude

As a verification of this fact, see *figure* 3.7. Two coördinate systems are shown, one rotated through an arbitrary angle α from the other. We demonstrate that the operation of finding the length of a given vector $\vec{V}$

produces the same result in both systems. (For clarity and ease of understanding, this is done in two dimensions).

From the diagram, the Pythagorean theorem applied to triangle **POQ** indicates that

$$|\vec{V}| = \sqrt{V_x^2 + V_y^2}$$

while applied to triangle **POQ'**, it indicates that

$$|\vec{V}| = \sqrt{V_{x'}^2 + V_{y'}^2}$$

The right side of the first expression is the magnitude as computed from the unprimed system: the right side of the second is the magnitude as computed from the primed system. The Pythagorean theorem ensures that both expressions produce the same scaler $|\vec{V}|$.

As an example of an illegitimate vector operation which produces a scaler, consider the operation of adding the components of a vector together. This operation performed on the vector $\vec{v} = \vec{\imath} + \vec{\jmath}$ produces the result **2** (= **1** + **1**). Now rotate the axes through **45°**, counterclockwise. The same vector expressed in the new system is $\vec{v} = \sqrt{2}\,\vec{\imath} + 0\vec{\jmath}$. Performing the same operation on the vector again, the result is $\sqrt{2}$ (= $\sqrt{2} + 0$). Therefore adding the components of a vector together to produce a scaler is not a legitimate vector operation: the result depends on the coördinate system used to express the given vector.

3.2.6.4 The Addition and Subtraction of Vectors

Material similar to what follows was presented earlier (See *section* 0.2.2).

Addition and subtraction of vectors are defined by the so-called parallelogram rule. Let $\vec{u}$ and $\vec{v}$ be two given vectors. Their sum $\vec{w} = (\vec{u} + \vec{v})$ is found by placing the tail of $\vec{v}$ at the head of $\vec{u}$; the sum $\vec{w}$ is the third side of the triangle so formed (this corresponds to the path *PQR*. See *figure* 3.8.). Notice that $\vec{w}$ is also the third side of the triangle formed if the tail of $\vec{u}$ is placed at the head of $\vec{v}$ (this corresponds to the path *PQ'R*). We therefore have

$$\vec{w} \;=\; \vec{u} \,+\, \vec{v} \;=\; \vec{v} \,+\, \vec{u}$$

which is to say that the addition of vectors is commutative. The sum $\vec{w}$ is one diagonal of the parallelogram whose sides are $\vec{u}$ and $\vec{v}$; the difference of the vectors, $\vec{w}' = \vec{u} - \vec{v}$ is the other diagonal with the tail of the difference vector at the head of the subtracted vector.

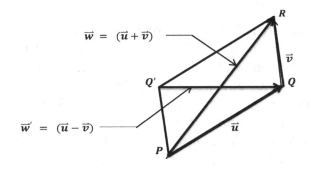

figure 3.8

The Legitimacy of the Parallelogram Rule

To check the consistency of these claims look at the diagram and verify that the vector $\vec{v}$ added to the vector $(\vec{u} - \vec{v})$ is equal to the vector $\vec{u}$. The above definitions for the addition and subtraction of vectors in no way reference a coördinate system and hence satisfy the legitimacy criterion cited above. Therefore

> *the parallelogram rule for the addition and subtraction of vectors is a legitimate vector operation.*

Note that subtraction can be defined as follows: take the negative of the subtracted vector by switching its head and tail and then add the result to the other addend (see *section* 0.2.2). This definition reduces subtraction to addition and therefore what is determined here about addition refers also to subtraction.

It is expedient at this point to illustrate that given two vectors and an arbitrary direction, the projection of their vector sum in the given direction equals the sum of their projections in that same direction. This result will be of use here and in the next section.

The following construction (see *figure* 3.9) makes the conclusion obvious. Construct three planes all normal to the line σ. Let one contain the point *P,* the second, *Q* and the third, *R*. Let *P'*, *Q'* and *R'* be the points where these planes intersect the line σ: then the projection of $\vec{u}$ in the direction of σ is the segment *P'Q'*, and similarly for the projections of $\vec{v}$ and $\vec{w}$. The projections of $\vec{u}, \vec{v}$ and $\vec{w}$ in the direction of σ (designated u_σ, v_σ and w_σ, respectively) are shown at the top of the diagram. It is clear from the figure that

$$u_\sigma + v_\sigma \;=\; w_\sigma.$$

Once again, the conclusion is

the sum of the projections of two given vectors in any direction is equal to the projection of their sum in that direction.

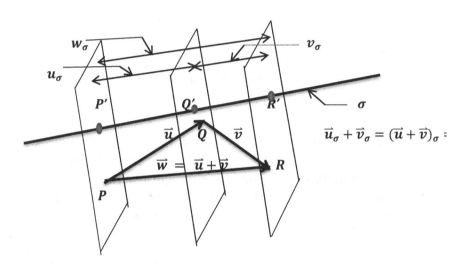

figure 3.9

The Sum of the Projections Equals the Projection of the Sum

Now assume that two vectors $\vec{u}$ and $\vec{v}$ are expressed in Cartesian coördinates and their sum (or difference) $\vec{w}$ is to be expressed in the same coordinates. We have, for the **x-**direction

$$w_x = (\vec{u} \pm \vec{v})_x$$

By the previous result, the projection in the **x-**direction of the sum (difference) on the right side of the above equation is the sum (difference) of the projections of the individual vectors. Therefore,

$$w_x \;=\; u_x \pm v_x.$$

The same holds for the **y**- and **z**-components. Hence, if

$$\vec{u} \;=\; u_x\vec{i} + u_y\vec{j} + u_z\vec{k}$$

$$\vec{v} \;=\; v_x\vec{i} + v_y\vec{j} + v_z\vec{k},$$

then

$$\vec{w} \;=\; \vec{u} \pm \vec{v}$$

$$\vec{w} \;=\; (u_x \pm v_x)\vec{i} + (u_y \pm v_y)\vec{j} + (u_z \pm v_z)\vec{k}$$

This is a simple rule: vectors in Cartesian coördinates are added and subtracted componentwise. The componentwise addition and subtraction of vectors extends to a sum with any number of addends. (The above result also extends to other coördinate systems.)

3.2.6.5 The Dot Product of Two Vectors

The first of two types of vector products to be presented here is called the dot product (aka the inner product, aka the scaler product). Its reason for being is its important role as a central element in the vector formulation of physics. The dot product of two vectors is a way of combining the vectors to produce a scaler quantity.

Let $\vec{u}$ and $\vec{v}$ be two given vectors. We make the following ...

> *Definition: The Dot Product (also called the Inner Product or Scaler Product) of $\vec{u}$ and $\vec{v}$ is the scaler quantity defined by the equation*

$$\vec{u} * \vec{v} = |\vec{u}||\vec{v}| \cos(\theta),$$

where the asterisk indicates the dot product and θ is the angle between the vectors.

The definition references only the vectors and the angle between them, so the dot product can be accepted as a legitimate vector operation (i.e. it does not reference any coordinate system).

Two things follow immediately from the definition:

1. **a vector dotted with itself is the square of its magnitude** $(cos(\theta) = 1)$

$$\vec{u} * \vec{u} = |\vec{u}||\vec{u}| \cos(\theta) = |\vec{u}|^2;$$

2. **the dot product of two non-zero vectors is zero if and only if the vectors are perpendicular** $(cos(\theta) = 0.)$

It is clear from the definition that the order of multiplication is inconsequential. Hence

$$\vec{u} * \vec{v} = \vec{v} * \vec{u},$$

that is to say, dot products commute.

We will also need the distributive law:

$$\vec{u} * (\vec{v} + \vec{w}) = \vec{u} * \vec{v} + \vec{u} * \vec{w}.$$

The following string of relations shows that the above equation is true:

$$\vec{u} * (\vec{v} + \vec{w}) = |\vec{u}||\vec{v} + \vec{w}| \cos(\theta)$$

$$\vec{u} * (\vec{v} + \vec{w}) = |\vec{u}|(\vec{v} + \vec{w})_u$$

where the subscript designates the component of the sum in parentheses in the direction of $\vec{u}$. By the result obtained above, the projection of the sum is the sum of the projections:

$$\vec{u} * (\vec{v} + \vec{w}) = |\vec{u}|(v_u + w_u)$$

$$\vec{u} * (\vec{v} + \vec{w}) = |\vec{u}|v_u + |\vec{u}|w_u$$

$$\vec{u} * (\vec{v} + \vec{w}) = |\vec{u}||\vec{v}| \cos(\alpha) + |\vec{u}||\vec{w}| \cos(\beta)$$

$$\vec{u} * (\vec{v} + \vec{w}) = \vec{u} * \vec{v} + \vec{u} * \vec{w},$$

where α is the angle between $\vec{u}$ and $\vec{v}$ and β is the angle between $\vec{u}$ and $\vec{w}$.

The fact that the dot product is distributive establishes that products can be found using the normal rules of algebra. Thus to find the dot product of vectors expressed in Cartesian coordinates, simply write them out and dot them using the multiplication rules of ordinary algebra:

$$\vec{u} * \vec{v} = \left(u_x \vec{\imath} + u_y \vec{\jmath} + u_z \vec{k}\right) * \left(v_x \vec{\imath} + v_y \vec{\jmath} + v_z \vec{k}\right).$$

The product of the two trinomials above has nine terms. Six of them are zero (because the dot product of two perpendicular vectors is zero):

$$\vec{\imath} * \vec{\jmath} = \vec{\imath} * \vec{k} = \vec{\jmath} * \vec{k} = 0$$

Also

$$\vec{\imath} * \vec{\imath} = \vec{\jmath} * \vec{\jmath} = \vec{k} * \vec{k} = 1$$

Using the above relations when multiplying the dot product out, the result is

$$\vec{u} * \vec{v} = u_x v_x + u_y v_y + u_z v_z.$$

Note that the last expression is a scaler and is easily found from the Cartesian coördinate representations of the vectors.

Exercise: Multiply the two trinomials above to obtain the previous expression.

3.2.6.6 The Cross Product of Two Vectors

The dot product of two vectors involves the projection of one of the vectors in the direction of the other. The cross product as you may have guessed involves the other component, the projection of one in the direction normal to the other. Unlike the dot product which is a scaler, however, the cross product is a vector.

> ***Definition:*** *The Cross Product (also known as the Vector Product) is defined as follows. Its magnitude is defined by*
>
> $$|\vec{u} \times \vec{v}| = |\vec{u}||\vec{v}| \sin(\theta)$$
>
> *and its direction is determined according to a right-hand rule. Consistency requires that the coordinate axes also be a right-handed system, i.e. the positive x direction crossed into the positive y direction is the positive z direction.*[10]

As in the case of the dot product, the definition references only the vectors and the angle between them and is therefore a legitimate vector operation.

Note: the cross product does ***not*** commute:

$$\vec{u} \times \vec{v} = -\vec{v} \times \vec{u}.$$

Exercise: verify the last relation by applying the right-hand rule to $\vec{u} \times \vec{v}$ and to $\vec{v} \times \vec{u}$.

---

[10] The right-hand rule: point the fingers of your right hand in the direction of the first vector of the cross-product. Curve the fingers slightly to point toward the second vector of the cross-product. If your fingers don't bend that way, flip your right hand over and start again. Your right thumb now points in the direction of the cross-product.

Happily, however, it **does** distribute. For proof, it must be shown that

$$\overrightarrow{w} \times (\overrightarrow{u} + \overrightarrow{v}) = \overrightarrow{w} \times \overrightarrow{u} + \overrightarrow{w} \times \overrightarrow{v}.$$

Proof: Refer to **figure 3.10**. For the sake of specificity, let the vector $\overrightarrow{w}$ point upward normal to a horizontal table top and consider two arbitrary vectors $\overrightarrow{u}$ and $\overrightarrow{v}$ and their sum $\overrightarrow{s} = (\overrightarrow{u} + \overrightarrow{v})$. The vectors $\overrightarrow{u}, \overrightarrow{v}$ and $\overrightarrow{s}$ form a triangle as shown. Think of this triangle as hovering above the table in some arbitrary orientation. Let α, β and γ be the angles between ($\overrightarrow{w}$ and $\overrightarrow{u}$), ($\overrightarrow{w}$ and $\overrightarrow{v}$) and ($\overrightarrow{w}$ and $\overrightarrow{s}$), respectively.

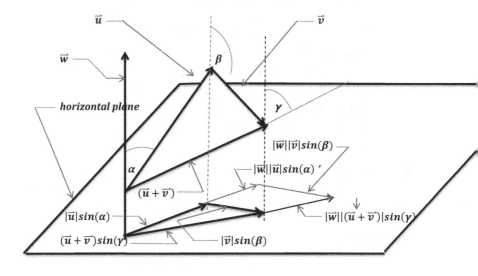

figure 3.10

The Distributivity of the Cross Product

There is no loss of generality by referencing the tail of $\overrightarrow{u}$ to a point along $\overrightarrow{w}$ as shown. The proof will be accomplished by noting the result of performing the following three operations on the triangle:

1. *project the triangle onto the table top;*
2. *scale the triangle by a factor of $|\vec{w}|$;*
3. *rotate the scaled projection 90^0.*

To simplify the analysis, consider only one of the vectors that make up the original triangle, say $\vec{u}$.

The projection of $\vec{u}$ is its component normal to $\vec{w}$ having magnitude $|\vec{u}|sin(\alpha)$. The scaled projection has magnitude $|\vec{w}||\vec{u}|\ sin(\alpha)$ which is the magnitude of the cross-product $|\vec{w} \times \vec{u}|$.

Let $\vec{U}$ be the rotated, scaled projection of $\vec{u}$. $\vec{U}$ is normal to $\vec{w}$ because it is in the plane of the table top. Also, $\vec{u}$ can be resolved into two components: one in the direction of $\vec{w}$, and one normal to $\vec{w}$. $\vec{U}$ is normal to both of these components: the first because it's normal to $\vec{w}$ (i.e. in the plane of the table top); the second because it was rotated 90^0 from the projection of $\vec{u}$.

If these results are collected, we have:

1. $\vec{U}$ is normal to both $\vec{w}$ and $\vec{u}$;
2. $|\vec{U}|$ = $|\vec{w} \times \vec{u}|$.

There are only two possibilities. They are

$$\vec{U} = \pm (\vec{w} \times \vec{u}).$$

This result applies to all three of the vectors $\vec{u}, \vec{v}$, and $\vec{s}$.

Now consider the original triangle which represents the vector relaton

$$\vec{s} = \vec{u} + \vec{v}.$$

This triangle stays intact under the operations described and the resulting triangle has sides

$$(\vec{w} \times \vec{s}), (\vec{w} \times \vec{u}) \text{ and } (\vec{w} \times \vec{v})$$

as was just shown. These cross-products are assembled into a triangle which represents the vector equation

$$\vec{w} \times \vec{s} = \vec{w} \times \vec{u} + \vec{w} \times \vec{v},$$

or $\quad \vec{w} \times (\vec{u} + \vec{v}) = \vec{w} \times \vec{u} + \vec{w} \times \vec{v}.$

But, this last relation for arbitrary $\vec{u}, \vec{v}$ and $\vec{w}$ is the distributive law. This completes the proof.

The result obtained will be needed to find an expression for the cross product of vectors when they are represented in a Cartesian coördinate system. Since the cross-product of a vector with a sum of vectors is the sum of the individual cross products, two vectors expressed in Cartesian coordinates can be multiplied out using the ordinary rules of algebra. Care must be taken however, not to reverse the order of multiplication of the unit vectors $\vec{i}, \vec{j}$ and $\vec{k}$. (Reversing the order changes the sign of their product.)

There is a mnemonic device to help remember how to cross one of the unit vectors into a second: write the unit vectors in (cyclic) order; then if you cross two adjacent unit vectors moving to the right, the result is the next one to the right; if you cross two adjacent unit vectors moving to the left, the result is the next one to the left with a minus sign.

$$\Rightarrow\Rightarrow\Rightarrow\Rightarrow \quad +++ \quad \Rightarrow\Rightarrow\Rightarrow\Rightarrow$$
$$\vec{i} \qquad \vec{j} \qquad \vec{k} \qquad \vec{i} \qquad \vec{j}$$
$$\Leftarrow\Leftarrow\Leftarrow\Leftarrow \quad --- \quad \Leftarrow\Leftarrow\Leftarrow\Leftarrow$$

Using these rules, we have, for example,

$$\vec{i} \times \vec{j} = \vec{k}; \qquad \vec{k} \times \vec{i} = \vec{j}; \qquad \vec{j} \times \vec{i} = -\vec{k}$$

The road has now been cleared to find an expression for the cross product of two vectors in Cartesian coordinates.

As before with the dot product, we simply write the vectors in Cartesian form and cross them:

$$\vec{u} \times \vec{v} = \left(u_x \vec{i} + u_y \vec{j} + u_z \vec{k}\right) \times \left(v_x \vec{i} + v_y \vec{j} + v_z \vec{k}\right)$$

Since the cross product is distributive, the above expression can be multiplied out according to the rules of normal algebra. Keeping in mind that any vector crossed with itself is zero, and using the above described rules for crossing the principle unit vectors, we have

$$\vec{u} \times \vec{v} = \left(u_y v_z - u_z v_y\right) \vec{i} + \left(u_z v_x - u_x v_z\right)\vec{j} + \left(u_x v_y - u_y v_x\right)\vec{k}.$$

Exercise: Show that this last expression is the same as the previous one. (Multiply the first one out using the rules outlined.)

3.2.7   A Splash of Vector Calculus

3.2.7.1 Outline

In the following section the differential formulas for the dot product and the cross product are derived. These things are on the border of the subject matter for this book. One of the formulas, the one in connection with the dot product, has application to the work-energy relation which will be derived here using vector notation. The remainder of this section which is simple and brief, is included for completeness.

3.2.7.2   Differential Formulas for the Dot Product and Cross Product

The differential formulas are obtained by applying the delta process of elementary calculus to the vector products. There is nothing different from the ordinary variable case but it is suggested that you read through the derivations with the knowledge that the variables are vectors so that you can convince yourself that the extension of these formulas to vector algebra is justified.

The formula for the differential of the dot product is

$$d(\vec{u} * \vec{v}) \;=\; \vec{u} * d\vec{v} \;+\; \vec{v} * d\vec{u}.$$

To obtain it, apply the delta process to the dot product:

$$d(\vec{u} * \vec{v}) \;=\; (\vec{u} + d\vec{u}) * (\vec{v} + d\vec{v}) \;-\; \vec{u} * \vec{v}$$

$$d(\vec{u} * \vec{v}) = \vec{u} * \vec{v} + \vec{u} * d\vec{v} + d\vec{u} * \vec{v} + d\vec{u} * d\vec{v} - \vec{u} * \vec{v}$$

The first and last terms cancel; the fourth term goes to zero (only first-order differentials survive the limiting process as the changes go to zero; the fourth term is a product of two first-order differentials which is a second-order differential). The result is

$$d(\vec{u} * \vec{v}) \;=\; \vec{u} * d\vec{v} \;+\; d\vec{u} * \vec{v}$$

$$d(\vec{u} * \vec{v}) \;=\; \vec{u} * d\vec{v} \;+\; \vec{v} * d\vec{u}.$$

This sequence of steps is exactly the same as those used to find the differential of the product of two scaler functions. Look it over to see that it extends easily to vectors without offending your sense of vector algebra.

The rule for the differential of the cross-product is derived in exactly the same way as for that of the dot product by replacing all dot product indicators (*) by cross-product indicators (×). The only exception is the omission of the last step because in the case of cross-products the order of the factors cannot be reversed without an accompanying change in sign. As an exercise, try to write it out. The result should be

$$d(\vec{u} \times \vec{v}) \;=\; \vec{u} \times d\vec{v} \;+\; d\vec{u} \times \vec{v},$$

or $\qquad d(\vec{u} \times \vec{v}) \ = \ \vec{u} \times d\vec{v} \ - \ \vec{v} \times d\vec{u}$.

If there is any problem, refer to the derivation for the dot product. Also note that if $f(x)$ and $g(x)$ are two scaler functions, the elementary calculus formula for the differential of their product is

$$d(fg) \ = \ fd(g) \ + \ g\,d(f)$$

This is essentially the same result as those obtained for the two types of vector products.

3.2.7.3 Vector Derivation of the Work-Energy Relation

Recall that it was determined by the use of a vector diagram (see *sections* $1.2.5.1$ and $1.2.5.2$) that for any given mass the component of force in the direction of motion was solely responsible for the change in speed. This motivated an analysis to answer the question, "What exactly is the quantitative relation between this component of force and the speed?" The answer was the work-energy law.

The previous expression for the differential of the dot product of two vectors can be used to arrive at this same result. To do this, consider the expression for the differential of the dot product in the special case that $\vec{u}$ and $\vec{v}$ are equal:

$$d(\vec{u} * \vec{v}) \ = \ \vec{u} * d\vec{v} \ + \ \vec{v} * d\vec{u}$$

$$d(\vec{u} * \vec{u}) \ = \ 2\vec{u} * d\vec{u}$$

$$\Longrightarrow \quad \tfrac{1}{2}d(|\vec{u}|^2) \ = \ \vec{u} * d\vec{u}$$

This last expression will be needed in a minute.

137

Let $d\vec{s}$ be a differential vector pointing in the direction of motion whose length is a differential length ds along the path. Then the differential amount of work dW done on the body is

$$dW \;=\; F_s ds \;=\; \vec{F} * d\vec{s}$$

$$dW \;=\; m\frac{d\vec{v}}{dt} * d\vec{s}$$

$$dW \;=\; m\frac{d\vec{s}}{dt} * d\vec{v}$$

$$dW \;=\; m\vec{v} * d\vec{v}.$$

Using the formula derived above,

$$dW \;=\; d\left(\frac{1}{2}m\,|\vec{v}|^2\right).$$

Integrating this last expression from A to B,

$$W_{AB} \;=\; \frac{1}{2}m\,|\vec{v}_B|^2 \;-\; \frac{1}{2}m\,|\vec{v}_A|^2$$

where W_{AB} is the work in going from A to B, v_A is the velocity at A and v_B is the velocity at B. (**Note:** $\vec{F} * d\vec{s} \;=\; |\vec{F}_s||d\vec{s}| \;=\; dW$.)

Compare this derivation of the work-energy relation to that of *section* 1. 2. 4. 2.)

3.2.8   Momentum and Kinetic Energy in the *CM* System

3.2.8.1 General Remarks

Since the origin of the *CM* coördinate system is the center of the mass distribution in the sense that the moments of mass around this point add to zero, it would be expected that the equations of motion for a system acquire a kind of simplifying symmetry when referred to this coördinate system. This is in fact the case and in preparation for exploiting this fact, the equations which relate the momentum and kinetic energy as measured in the laboratory coordinates to those as measured in the *CM* system are derived in this section.

In the subsequent section, the problems of inelastic and elastic collisions will be reconsidered by setting up the equations in the *CM* system. Comparison of the equations and their solutions from this viewpoint with those of *sections* **3.2.2** and **3.2.3** will illustrate the degree of simplification afforded by this approach.

3.2.8.2   The Momentum Measured from an Arbitrary System

In this section and the next, the unprimed quantities refer to any arbitrary coördinate system and the primed quantities refer to the *CM* system. Note that there are no assumptions concerning the inertial or non-inertial nature of the *CM* system: the cases considered are completely general.

Let O be the origin of the laboratory system, and $\vec{r}_i$ the vector from O to the ith mass; let O' and $\vec{r}_i'$ be defined analogously in the *CM* system; let $\vec{R}$ be the vector from O to O' (refer to *figure* **3.11**). Then

$$\vec{r}_i = \vec{R} + \vec{r}_i'$$

Multiply this relation by the ith mass m_i and add the result over all the masses:

$$\sum_i m_i \vec{r}_i = \sum_i m_i \vec{R} + \sum_i m_i \vec{r}_i'$$

Note that the last sum computes the location of the **CM** in the **CM** system: it's at the origin because that's the definition of a **CM** system. Hence, the last term is zero. The vector $\vec{R}$ is constant and factors out of the second sum and the sum of the masses is **M**.

$$\sum_i m_i \vec{r}_i = MR$$

Take the time-derivative of the last equation:

$$\sum_i m_i \frac{d\vec{r}_i}{dt} = M\frac{d\vec{R}}{dt}$$

$$\sum_i m_i \vec{v}_i = M\vec{V}$$

where **M** is the total mass of the system and $\vec{V}$ is the velocity of the **CM**. In words,

> *the total momentum of a system as seen from any arbitrary coördinate system is equal to the momentum of the CM as seen from that system.*

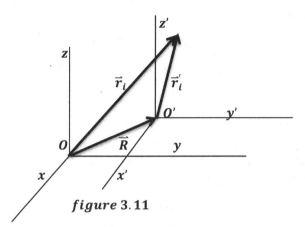

figure 3.11

The Momentum in an Arbitrary Coördinate System

In the particular case that the arbitrary system is the **CM** system itself, then the **CM** is positioned at the origin and stays there, so **V = 0.** It can be concluded that

$$\sum_i m_i v'_i = 0.$$

Hence,

the total momentum in the CM system is always zero

3.2.8.3 The Kinetic Energy Measured from an Arbitrary System

A similar approach is used to relate the total kinetic energy as seen from an arbitrary system to that as seen from the **CM** system.

First consider the ith mass and as before start with its position vector:

$$\vec{r}_i \;=\; \vec{R} \;+\; \vec{r}'_i \,;$$

take the time-derivative:

$$\vec{v}_i \;=\; \vec{V} \;+\; \vec{v}'_i \,.$$

Since we want to relate the kinetic energies, we need an expression for the square of the speed. To get such an expression, recall from the remarks about dot products that for any vector

$$\vec{w} * \vec{w} \;=\; |\vec{w}|^2$$

The squares of the speeds can be related by dotting the velocity equation with itself and using the above relation:

$$|\vec{v}_i|^2 \;=\; \left(\vec{V} + \vec{v}'_i\right) * \left(\vec{V} + \vec{v}'_i\right),$$

$$|\vec{v}_i|^2 \;=\; \left|\vec{V}\right|^2 \;+\; 2\,\vec{V} * \vec{v}'_i \;+\; |\vec{v}'_i|^2 \,.$$

The kinetic energies can now be related: multiply the last equation by $\left(\frac{1}{2} m_i\right)$. The result is

$$\frac{1}{2}\, m_i |\vec{v}_i|^2 \;=\; \frac{1}{2}\, m_i \left|\vec{V}\right|^2 \;+\; m_i\, \vec{V} * \vec{v}'_i \;+\; \frac{1}{2}\, m_i |\vec{v}'_i|^2 \,.$$

You have a relation like this one for each mass in the system. The result sought concerns the **total** energies involved. It can be found by adding these equations together over all the masses. You then get

$$\sum_i \frac{1}{2}\, m_i |v_i|^2 \;=\; \sum_i \frac{1}{2}\, m_i |V|^2 \;+\; \sum_i m_i\, \vec{V} * \vec{v}'_i \;+\; \sum_i \frac{1}{2} m_i |\vec{v}'_i|^2 \,.$$

It looks like a hopeless mess, but don't despair. It works out quite nicely.

Consider each term separately:

1. **the term on the left is the sum of all the kinetic energies in the system and equals the total energy T of the system;**

2. **in the first term on the right, there is a constant factor of $\left(\frac{1}{2}|V|^2\right)$ in each term which can be taken out of the summation. What is left is the sum of the m_i's which is just the total mass M of the system;**

3. **the second term on the right has a constant vector $\tilde{\vec{V}}$ dotted with all the (v_i')'s. But the dot product is distributive so dotting $\vec{V}$ with each $\vec{v}_i$ and adding them together is the same as adding the (v_I)'s first and dotting $\vec{V}$ with the whole sum. But the sum is just the total momentum in the CM system and it was just shown above that that is always zero;**

4. **the last term is just the total kinetic energy T' as seen in the CM system.**

Hence the expression above is reduced as follows

$$\sum_i \frac{1}{2} m_i |\vec{v}_i|^2 \;=\; \sum_i \frac{1}{2} m_i |\vec{V}|^2 \;+\; \sum_i m_i \vec{V} * \vec{v}_i' \;+\; \sum_i \frac{1}{2} m_i |\vec{v}_i'|^2$$

$$\sum_i \frac{1}{2} m_i |\vec{v}_i|^2 \;=\; \frac{1}{2} |\vec{V}|^2 \sum_i m_i \;+\; \vec{V} * \sum_i m_i \vec{v}_i' \;+\; \sum_i \frac{1}{2} m_i |\vec{v}_i'|^2$$

$$T \;=\; \frac{1}{2} M |\vec{V}|^2 \;+\; 0 \;+\; T'$$

$$T = \frac{1}{2} M|\vec{V}|^2 + T'.$$

The result in words is

> *the total kinetic energy of a system measured from an arbitrary coördinate system is the kinetic energy of the CM system plus the kinetic energy in the CM system.*

Since kinetic energy is always non-negative, this last relation indicates that the kinetic energy as measured in the *CM* system is less than that measured from any other system.

3.2.9 Collisions in the *CM* System

3.2.9.1 Inelastic Collisions

The problem of the inelastic collision of two bodies presented in *section* 3.2.2 is now reconsidered as it is seen from the *CM* system. The given conditions were these: a mass m_1 is moving with speed v_1 along the negative x-axis toward the origin where a mass m_2 is initially at rest; upon impact the masses adhere to each other. What is the final state of the system?

In the following two sections, the $v's$ are velocities in the laboratory system; the $w's$ are velocities in the *CM* system; unprimed quantities are values before collision; primed quantities, after collision.

Except for finding the velocity $\vec{V}$ of the *CM*, this problem can be solved by inspection. For the moment suppose that $\vec{V}$ is known. Then the solution is as follows. From the *CM* system the momentum is zero (as it always is in the *CM* system with both masses moving toward the origin where they collide. Since they adhere to each other and since the final momentum must also be zero (momentum is conserved), they remain at the origin. Hence, there is no motion after the collision in the *CM* system. In the laboratory system, the masses move in the positive *x*-direction with velocity $\vec{V}$. **work and   30-35, 43 work and 30-35, 43**

Although this is the solution (except for finding $\vec{V}$), the calculations are carried through as an illustration of the method for analyzing a problem from the *CM* system. The following three steps define the method:

1. *translate the problem from the laboratory system to the CM system;*
2. *write the equations in the CM system and solve them;*
3. *translate the results back to the laboratory coordinates.*

If P_i and P_f are respectively the initial and final momenta, and V is the speed of the *CM*, then

$$P_f \;=\; P_i$$

$$(m_1 + m_2)\,V \;=\; m_1 v_1$$

and the speed of the *CM* is found to be

$$V \;=\; \frac{m_1 v_1}{m_1 + m_2}$$

The speeds in the *CM* system are:

$$w_1 \;=\; v_1 - V$$

$$w_1 \;=\; \frac{m_2 v_1}{m_1 + m_2}$$

145

and
$$w_2 = -V$$
$$w_2 = -\frac{m_1 v_1}{m_1 + m_2}$$

(The total momentum in the *CM* system should be zero. As an exercise, verify this.)

Let P'_f be the final momentum in the *CM* system.

$$P'_f = 0$$
$$(m_1 + m_2)\, w' = 0$$
$$\Rightarrow \quad w' = 0$$

Note that the problem was solved by inspection. The three-step process which is overly elaborate for this problem was presented as an illustration of the method. Attention should be focused on the steps and how they are applied.

As a final observation, recall that the energy in the laboratory system is related to the energy in the *CM* system by

$$T = \frac{1}{2} M |\vec{V}|^2 + T'.$$

After the collision T' the (kinetic) energy in the *CM* system is zero which is its minimum value. Since *M* and *V* (in the first term on the right) cannot change, *T* the energy in the laboratory system is also at its minimum value. This indicates that

> *the kinetic energy loss in a perfectly inelastic collision is greater than that lost in any other type of collision.*

3.2.9.2 Elastic Collisions

This problem can also be solved by inspection when considered from the *CM* system. The pertinent information is this: the bodies collide at the origin and their *CM* must remain there. This implies that they must fly apart in opposite directions while maintaining the proper moment arms so the *CM* does not move. In order to maintain the proper moment arms around the origin, the ratio of the velocities must be the same after collision as before, i.e.

$$\frac{w_1}{w_2} = \frac{w_1'}{w_2'}$$

If T and T' represent the initial and final kinetic energies, then:

$$w_1' > w_1 \implies w_2' > w_2 \text{ and } T' > T.$$

This is impossible since energy is conserved and hence cannot be gained in the process;

$$w_1' < w_1 \implies w_2' < w_2 \text{ and } T' < T$$

which implies that energy has been lost. This is impossible for the same reason. Therefore,

$$w_1' = w_1 \text{ and } w_2' = w_2,$$

i.e. each mass leaves the crash site with the same speed as it had before collision.

In addition to the fact that the analysis can be done by inspection, we have already gotten more information than was afforded by all the algebraic antics of *section* 3.2.3. Recall that there were three equations in four unknowns when we set this up so there was insufficient information to find a complete solution. The one unknown left undetermined by the above is the angle between the *x*-axis and the line of motion after collision; all other information concerning the physics has been determined. The only task left is a mathematical one --- to translate back into the laboratory coordinates.

Nevertheless, the three-step process for using the **CM** system to analyze the problem will be carried through as before. (Refer to *figure* **3.1.1**.)

Step 1:   *Translate the problem into the CM system.*

As in the case of the inelastic collision analysis, the conditions before impact are:

$$V = \frac{m_1 v_1}{m_1 + m_2}$$

$$w_1 = |v_1 - V| = \frac{m_2 v_1}{m_1 + m_2}$$

$$w_2 = |v_2 - V| = \frac{m_1 v_1}{m_1 + m_2}$$

Step 2:   *Write the equations of motion in the CM system and solve them.*

The following equations are consistent with the diagram of *figure* **3.12** and are written in such a way as to let the angles take care of the signs.

Conservation of momentum

x component:

$$m_1 w_1' \cos(\theta) + m_2 w_2' \cos(\varphi) = 0 \qquad (1)$$

y component:

$$m_1 w_1' \sin(\theta) + m_2 w_2' \sin(\varphi) = 0 \qquad (2)$$

Conservation of energy

$$\frac{1}{2}m_1v_1^2 + \frac{1}{2}m_2 \cdot v_2^2 = \frac{1}{2}m_1 v'^2_1 + \frac{1}{2}m_2 \cdot v'^2_2 \qquad (3)$$

Equations **(1)** and **(2)** can be rewritten:

$$m_1 w'_1 \cos(\theta) \quad = \quad -m_2 w'_2 \cos(\varphi) \qquad (1')$$

$$m_1 w'_1 \sin(\theta) \quad = \quad -m_2 w'_2 \sin(\varphi) \qquad (2')$$

Dividing **(2')** by **(1')**, the result is

$$tan(\theta) \quad = \quad tan(\varphi)$$

with solutions

$$\varphi = \theta \quad and \quad \varphi = (\pi + \theta)$$

The first of these solutions refers to the perfectly inelastic case when the masses adhere to each other upon collision. We are interested here in the second solution. In this case, the solution indicates that the masses fly apart in opposite directions; the momentum equations **(1')** and **(2')** each yield

$$m_1 w'_1 \quad = \quad m_2 w'_2 \qquad (4)$$

Reference back to the 2nd and 3rd equations of step 1 shows that the above speed ratio is the same as it was before collision.

Note that all the information gotten so far was obtained from the momentum equations; the energy equation has not yet been used. It should tell us that

$$w'_1 \quad = \quad w_1$$

$$w'_2 \quad = \quad w_2$$

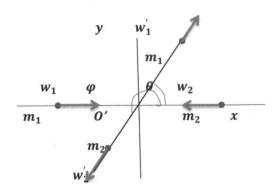

$$figure\ 3.12$$

Elastic Colisions in a CM System

The initial energy of the system was

$$T \;=\; \frac{1}{2}\,m_1 w_1^2 \;+\; \frac{1}{2}\,m_2 w_2^2$$

Substitute the expressions for w_1 and w_2 from step **1**:

$$T \;=\; \frac{1}{2}\,\frac{m_1 m_2^2 \;+\; m_1^2 m_2}{(m_1 \;+\; m_2)^2}\,v_1^2$$

$$T \;=\; \frac{1}{2}\,\frac{m_1 m_2}{m_1 \;+\; m_2}\,v_1^2$$

Apply the conservation of energy:

$$T_f \;=\; T_i$$

$$\frac{m_1 m_2}{m_1 \;+\; m_2}\,v_1^2 \;=\; m_1 w_1'^2 \;+\; m_2 w_2'^2$$

Eliminate w_2' using eq. (**4**) and fill in the missing algebra to obtain

$$w_1' = w_1$$

Also do the algebra to obtain

$$w_2' = w_1$$

At this point, everything that has been determined by inspection has been formally verified. The only task left is to refer the results back to the laboratory system. That is the business of ...

Step 3: *Translate the results back into the original coordinates.*

The translation back into the original coordinates is accomplished by adding the vector velocity of the *CM* to the velocities found above. Then the final velocities and the angle between them can be found in terms of the angle *θ.*

From the diagram in *figure* 3.12, it can be seen that the values of v_1' and v_2' given in terms of their *x*- and *y*- components are

$$v_{1_x}' = V + \frac{m_2 v_1}{m_1 + m_2} \cos(\theta)$$

$$v_{1_y}' = \frac{m_2 v_1}{m_1 + m_2} \sin(\theta)$$

$$v_{2_x}' = V - \frac{m_1 v_1}{m_1 + m_2} cos(\theta)$$

$$v_{2_y}' = \frac{m_1 v_1}{m_1 + m_2} sin(\theta)$$

This is the best answer that can be obtained from the information given, the final velocities in terms of the undetermined angle θ. (The incompleteness of the answer is the result of having only three equations in four unknowns.) It is proposed to find the angle $(\alpha + \beta)$ between the final velocities in the laboratory system. You will be given a good head start on this problem and the rest will be left as an exercise.

Note first that the resolution of the trig/algebra problem implied by the diagram of *figure* 3.12 is nothing short of a nightmare: an attempt to extract the desired information by ordinary methods would most likely fizzle by the time you find yourself up to your knees in symbols and equations which don't look like they're leading you anywhere. In the case at hand, the cavalry that comes over the hill just in the nick of time to save the day arrives in the attire of vector algebra. There are two tricks that can be used which bring the finding of an expression for $(\alpha + \beta)$ into the realm of possibility.

The first of these is the use of the dot product:

$$\vec{v}_1' * \vec{v}_2' = |\vec{v}_1'||\vec{v}_2'| \, cos(\alpha + \beta)$$

$$\Rightarrow \quad cos(\alpha + \beta) = \frac{\vec{v}_1' * \vec{v}_2'}{|\vec{v}_1'||\vec{v}_2'|}$$

This would be enough to come up with the answer, but notice that the magnitudes in the denominator of the last equation are messy: they involve the square root of the sum of the squares of the components for each vector. This situation can be considerably improved by using the second trick (maybe you guessed it already), the use of the cross product.

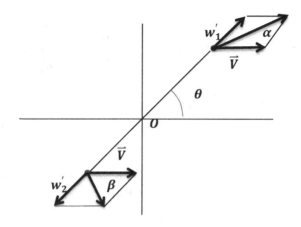

$$figure\ 3.13$$

Translating Back to Laboratory Coördinate

Note that

$$|\vec{v}_1' \times \vec{v}_2'| \quad = \quad |\vec{v}_1'||\vec{v}_2'|\ sin(\alpha + \beta)$$

$$\implies \quad sin(\alpha + \beta) \quad = \quad \frac{|\vec{v}_1' \times \vec{v}_2'|}{|\vec{v}_1'||\vec{v}_2'|}$$

If the second equation is divided by the first, the messy magnitudes cancel and

$$tan(\alpha + \beta) \quad = \quad \frac{|\vec{v}_1' \times \vec{v}_2'|}{\vec{v}_1' * \vec{v}_2'}.$$

In *sections* **3. 2. 6. 5** and **3. 2. 6. 6**, formulas were derived expressing the dot and cross products in terms of their Cartesian coordinates. That's your head start.

Exercise: Use the formulas referenced and the results obtained above to find an expression for $tan(\alpha + \beta)$. You should get

$$tan(\alpha + \beta) = \frac{v_1 V sin(\theta)}{V^2 - \dfrac{m_1 - m_2}{m_1 + m_2} V v_1 cos(\theta) - \dfrac{m_1 m_2}{(m_1 + m_2)^2} v_1^2}.$$

We will particularize the problem at this point by assuming the masses to be equal. This is the case for which the angle between the final velocities is **90°** (see *section* **3.2.3**). This result will now be obtained again.

Let $m = m_1 = m_2$. Then the equations for the final velocities in the laboratory system become:

$$v'_{1_x} = \frac{v_1}{2} [1 + cos(\theta)];$$

$$v'_{1_y} = \frac{v_1}{2} sin(\theta)$$

$$v'_{2_x} = \frac{v_1}{2} [1 - cos(\theta)]$$

$$v'_{2_y} = -\frac{v_1}{2} sin(\theta)$$

The fact that the angle between these two vectors is **90°** will be shown in two ways:

1. **by straightforward trig;**

2. **by showing their dot product is zero.**

Method 1:   *straightforward trig*

$$tan(\alpha) = \frac{v'_{1_y}}{v'_{1_x}} = \frac{sin(\theta)}{1 + cos(\theta)}$$

$$tan(\beta) = -\frac{v'_{2_y}}{v'_{2_x}} = \frac{sin(\theta)}{1 - cos(\theta)}$$

The formula for the tangent of the sum of two angles is

$$tan(\alpha + \beta) = \frac{tan(\alpha) + tan(\beta)}{1 - tan(\alpha)\,tan(\beta)}$$

Using the above expressions,

$$tan(\alpha + \beta) = \frac{\dfrac{sin(\theta)}{1 + cos(\theta)} + \dfrac{sin(\theta)}{1 - cos(\theta)}}{1 - \dfrac{sin^2(\theta)}{1 - cos^2(\theta)}}.$$

Clear the complex fractions:

$$tan(\alpha + \beta) = \frac{2sin(\theta)}{1 - cos^2(\theta) - sin^2(\theta)}$$

$$tan(\alpha + \beta) = \frac{2sin(\theta)}{1 - 1} = \infty$$

$$\Rightarrow \quad \alpha + \beta = 90°$$

Method 2: *show the dot product of the final vectors is zero.*

$$v_1' * v_2' = [1 + cos(\theta)][1 - cos(\theta)] - sin^2(\theta)$$

$$v_1' * v_2' = 1 - cos^2(\theta) - sin^2(\theta)$$

$$v_1' * v_2' = 1 - 1 = 0$$

$$\Rightarrow \quad \alpha + \beta = 90°$$

Using the dot product is clearly much easier than going the trig/algebra route.

Thus all the results obtained in *section 3.2.3* have now been obtained again by viewing the collision from the *CM* system. The fact that the problem can be solved by inspection in the *CM* system bespeaks the insights and simplifications afforded by this approach. It is also worthy of note that when viewed from the laboratory system in *section* 3.2.3, the equations did not indicate a clear path to the general result for the angle between the final velocities. The analysis from the *CM* system did although the details were not presented here (see *section* 3.2.8.2). Furthermore, the current analysis indicates that the physics involved has a simple straightforward clarity about it: the difficulty in getting the solution resides in the trigonometry required to transfer the description out of the *CM* system back into the laboratory system.

A good deal of theoretical machinery was invoked to present a rather complete description of collision phenomena. In addition to the fact that it is important to learn the details of applying this machinery, collision problems deserve considerable attention because they have widespread application not only in mechanics but also in celestial mechanics and atomic physics.

One could easily imagine mechanical situations in which a perfectly inelastic collision occurs, for example, a dart hitting and sticking into a moving object. In atomic physics, whenever an atom captures an electron or absorbs a photon, it recoils according to the equations of a perfectly inelastic collision. The statement that a collision is perfectly inelastic is therefore not necessarily an idealization in the context of either mechanics or atomic physics.

A perfectly elastic collision on the other hand, is not so easily realizable in mechanics: there is always some friction or deformation when bodies come into contact. However, there are some materials like spring steel which come close to the definition of perfectly elastic.

Note however, that the word 'collision' need not be restricted to bodies which actually come into contact with each other. An example is the near collision of two celestial bodies. When they are in proximity, they exert a gravitational attraction on each other and if they manage to escape mutual capture (i.e., they do not end up orbiting around each other), they déflect each other's path. Since the gravitational field is conservative, the final paths of the bodies (after they are free of each other's gravitational attraction) are determined by the equations of a perfectly elastic collision. (Note that if they do end up orbiting each other, then the equations of a perfectly inelastic collision apply.)

In atomic physics, perfectly elastic collisions abound and occur similarly to the previous example except that the forces are more often electro-dynamic rather than gravitational. They arise most often from the attraction and repulsion of electric charges and as in the gravitational case, the force fields are conservative. The deflection of an electron beam by a charged body is an example. Again the bodies never come into contact (in fact the concept of 'in contact' does not withstand scrutiny in any context, even if we're talking about billiard balls).

Thus when the concept of collisions is extended to situations like these, it becomes clear that neither 'perfectly elastic' nor 'perfectly inelastic' is an idealization in mechanics, celestial mechanics or atomic physics.

Collision phenomena of both of these extreme types were very important in the development of modern physics. For almost a century now, the design of many measuring devices has been centered on forming a beam of some particular kind of particle to direct at some object of study, like an electron beam at a crystal. These beams have been used extensively to discern the qualities of the study object from the deflections it affected. Such studies have been used extensively in spectroscopy, crystallography, laser optics and quantum mechanics. These are a few of the many investigations that find their basis in the analysis of collisions.

3.3   MOTION IN A CONSTANT FORCE FIELD

3.3.1   Force Fields

3.3.1.1 General Remarks about Fields

When one deals in questions of theoretical physics, he is often called upon to walk the tightrope that connects metaphysics with 'what seems to be so'. His theories must have logical consistency and be based on the evidential results of repeatable experiments. This is the discipline of western science.

In this regard, there is a basic question which arises concerning what is meant by a field. Although this question is to be addressed only briefly in the context of this book, the use of the concept of a force field demands that prior to its use, the concept be positioned correctly in the fabric of the intricately woven ideas we call physics.

The question arises because a force field does not satisfy the basic requirement of western science:

> *no one has ever seen, isolated or measured a force field. How then, can we accept its existence as a viable notion?*

And how is it that the notion has accrued such solidity that today we speak freely of gravitational fields and electromagnetic fields as though their existence were beyond question?

An example will serve to answer these questions and to clarify what is meant by 'no one has ever measured a force field', but first, a few remarks about the classification of observable forces when physics was an adolescent science. It was observed early on that whenever two bodies came into contact, each could exert a push or a pull on the other and in so doing, alter the other's state of motion. There was no problem accepting the nature of these contact forces and their relation to movement because they were so close to everyday experience.

However, the early theories of universal gravitation introduced a different kind of force: an object could exert a push or a pull on another even when they didn't touch. The force was categorized as an *'action-at-a-distance'* force. There seemed to be no a priori reason why the existence of one celestial body should affect the motion of another when it wasn't even touching it. Forces acting at a distance were not as readily accepted as were contact forces: they were alien to everyday experience, and while they were intriguing and seemed to guard some great secrets of the universe, they also needed explanation.

Now anyone who studies the nature of so-called contact forces on an atomic level becomes aware that they are just as mysterious and intriguing because a magnified picture of a point of contact reveals that the forces which are in play during the 'contact' also act across a distance even though that distance is of atomic dimensions. The actual concept of contact between two bodies eludes a strict definition. Nevertheless, this classification of forces into contact forces and action-at-a-distance forces, although slightly erroneous, spurred investigators into considering the mysteries of universal gravitation and later, the mysteries of electrical forces.

Field theory was only one of the attempts to address the issue of forces which act at a distance. Three physicists, Maxwell, Faraday and Hertz developed the first field theory during the latter half of the 1800's. But this is jumping ahead: we return now to the time of Newton and trace the thinking concerning action at a distance which was in vogue and how it changed as time went on.

The gravitational force will be used as an example, using it as a central issue in this regard. Note however that electrostatic forces follow a law which is similar to the law of universal gravitational attraction in that the force of attraction is inversely proportional to the square of the distance between the interacting bodies. Gravitational and electrostatic forces have fundamental differences also, but these are not pertinent to the following discussion. Therefore, everything stated in this context concerning gravitational forces applies equally well to electrostatic forces as long as the masses are replaced with charges (with their signs) and the proportionality constant is changed appropriately.

3.3.1.2 Newton's Corpuscular Theory of Gravitation

Newton's law of universal gravitation says that any two masses exert a force of attraction on each other which is proportional directly to the product of their masses and inversely to the square of the distance between them:

$$F \quad = \quad -G\frac{m_1 m_2}{r^2},$$

where G is the universal gravity constant and the minus sign indicates that the force is attractive. He implied this from the motions of celestial bodies from which he ascertained that if $F = ma$, then they move as if acted upon by this force. There is no problem with this because this effect can and has been produced in a laboratory by measuring the angular displacement of a dumbbell suspended from a delicate torsion pendulum when a large mass is brought into proximity. So this law is verifiable by a repeatable experiment. We are not talking about dynamics now: dynamics relates the forces to the motion. We are talking about the *form* of this strange force, the force of gravity.

Newton wanted to explain why masses should universally attract one another. He asked the question, 'what mechanical business do two distant masses have with each other?' As described in the previous section, contact forces were more readily accepted as physical realities at that time than were action-at-a-distance forces. In keeping with this, he devised a corpuscular theory which reduces action-at-a-distance forces to contact forces.

According to this theory, the existence of a mass is accompanied by a continuous stream of small corpuscles which proceed from outer space and converge on the mass. With this postulate, the force is explained by the fact that a second mass introduced into the vicinity of the first would be in the midst of the corpuscular stream and as a result of the bombardment would experience a constant push in the direction of the first body. If the effect of each corpuscle were constant, the net force would be proportional to the rate of incidence, i.e., to the number per second of corpuscles incident on the second body.

The appeal of this theory comes from the fact that the inverse square field is precisely the one necessary to produce the correct dependence of the gravitational force on the distance between the bodies. Any other dependence would require a more complicated postulate involving the creation or destruction of some corpuscles at a rate that would depend on distance from the body, or some variation in the individual impact which would also depend on distance from the body.

To see this, consider the second body at some distance R from the first body and let f be the force of attraction. Imagine a sphere of radius R around the first body. Looking at the system from outer space, it can be seen that the force experienced by the second body is proportional to its projected area onto the sphere. Let s be the flux per unit area on the surface of the sphere. If the second body is now moved to a distance $2R$, and a sphere of radius $2R$ is now imagined, the projected area of the 2nd body is the same as before while the flux[11] of corpuscles per unit area would be one-fourth of its earlier value because they are spread out over 4 times the area than they were in the case of the smaller sphere. The force on the second body would therefore be one-fourth the original force. The rule is 'double the distance, quarter the force'. This is an inverse square dependence.

To be clear about the logic, there is no problem when two masses are present. Then the gravitational force is a measurable quantity and the law of universal gravitational attraction can be verified. But Newton postulated that the

---

[11] 'Flux' is the amount of flow across an area. There is a detailed discussion in *section 6.3.4* and its sub-sections.

existence of a single mass in space predisposes the space to have an effect on any other mass that **may** come into it.

The actual nature of this disposition is the postulated stream of corpuscles. If these corpuscles are not discernible in any context other than the gravitational attraction, then it is not possible to determine experimentally whether or not they exist. This is because the only way to attempt an experimental test of their existence would be to involve another mass, but that defeats the purpose since the experiment would simply measure the gravitational attraction of the two masses which is beside the point in this context: it would not answer the question concerning the pre-disposition of the space when there is only one mass and hence would not verify the existence or non-existence of the corpuscles.

Do gravitational corpuscles exist? No one knows. They have never been isolated or observed nor have any of their properties been determined. They therefore remain outside the realm of objects that physics can legitimately grace as having the property of existence.

Newton's position was to treat them as real entities because gravity forces behaved **as if** they were there, at least in the context of his own investigations. Eventually the corpuscle theory became bogged down with too many complexities and required too many additional assumptions to avoid non-sensical results. Had this not been the case, it would probably have the same status that today's field theories have.

As will be discussed in the next section, the force field of a single mass (or charge) cannot, in the same sense as described here be observed or measured. Consequently, fields are not admissible as realities according to the discipline of western science. By virtue of their widespread success, they are believed to exist by **implication,** not by **observation,** with the reason for this acceptance being that many things behave **as if** they were there. Know however, that this is not sufficient to establish them as realities. It is important to realize that the day may come when some more successful abstraction will supplant them.

3.3.1.3   Maxwell's Field Theory

For several decades starting in the 1850's, Maxwell, Faraday and Hertz made contributions which created the first field theory. The four Maxwell equations explain electrodynamics with the same comprehension as Newton's laws explain the dynamics of masses. The law they were studying governed the attractive (or repulsive) force between electrical charges. Like the gravitational attraction, this law can be and has been verified in the laboratory. It also has an inverse square dependence on distance:

$$F \;=\; K\,\frac{q_1 q_2}{r^2}$$

where K is a constant, the $q's$ are charges and r is the distance between them. Once again, the inverse square dependence on distance suggested that the space in the vicinity of a single charge is somehow pre-disposed by its presence. Instead of a corpuscular explanation for this action-at-a-distance force, it was said that a single charge q pre-disposes the space around itself by producing an electric field of magnitude

$$E \;=\; K\,\frac{q}{r^2}.$$

With this postulation, the force acting on a second charge *q'* introduced into the region is

$$F \;=\; Eq'$$

The field so defined is as unobservable as Newton's corpuscles: if one wished to measure the field, he would do so by introducing a charge into the region and once again, this is beside the point: such an experiment would measure only the *force* acting on the newly introduced charge and ***not the pre-disposition of the space before its introduction.***

The abstraction we call a field is the logical equivalent of Newton's corpuscles: things behave *as if* it were there. Field theory was more successful than corpuscle theory however, with the result that fields have become so integrated into the language of physics that it is easy to forget that they do not have the status of realities. They are a mathematically convenient abstraction which could possibly be replaced some day by a better one, or preferably by some theory that lends itself to experimental verification.

The success of field theory brought with it a metamorphosis in the concept of gravitation. The corpuscle theory died a slow death leaving no offspring to explain why masses should attract each other. The universal law of attraction was simply accepted as an experimental fact. When Maxwell's field theory came on the scene, the gravity forces were thought of as the effect of a gravitational field which was defined analogously to the electrostatic field:

$$U \;=\; -G\frac{m}{r^2}$$

where U is the field and m, the mass that generates it. Then, the force acting on a mass m' introduced into the field is:

$$F \quad = \quad Um'$$

All this is the exact analogue of the electrostatic case.

Whether or not electrostatic and gravitational fields exist, there are enough observable phenomena that behave as if they did, so that they will be used when investigating both local and universal gravitation. (There will be more about this in *sections* **6.3.1** and **6.3.2**.)

Note that at or near the surface of the earth, the gravitational field is

$$U \quad = \quad -G\,\frac{M}{R^2}$$

where M is the mass of the earth and R, its radius. When this quantity is evaluated, the result is g, the acceleration due to gravity. Within limits of height variation, it is fairly constant and is referred to as the local gravitational field. The attractive force between a mass m and the earth near the earth's surface is what we call the weight mg of the mass.

3.3.1.4 Einstein's General Relativity

General relativity is the most satisfying theory of gravitational attraction to be developed. In it, the mysterious and intriguing action-at-at-distance is explained by a distortion of space into a curved manifold in which Euclidean geometry no longer applies. This distortion is a function of the masses present.

General relativity has three salient features:

1. *there are no abstractions such as corpuscles or fields produced by one body into which a second body is introduced. The space curvature is a function of all the bodies present so that a second body cannot be considered to be an outside factor introduced into the space pre-conditioned by the first. This adheres closely to the experimental fact that the only measurable quantity is the force which requires the presence of both bodies;*

2. *only the distortion of the space away from Euclidean is necessary. After this is found, there is no mysterious force to be considered: the bodies simply move inertially in the curved space. Thus, both the force and the motion are explained by the single factor, space distortion;*

3. *it is theoretically possible to measure the disposition of the space in the presence of the masses: the curvature of three dimensional space can be measured FROM INSIDE THE SPACE. Hence the general theory of relativity does not postulate any feature or object whose existence or non-existence cannot be verified by experiment.*

To my knowledge, the actual **direct** measurement of the space curvature in the presence of masses has never been carried out: but the very fact that it can be measured **theoretically** (whether or not this has actually been done) means that general relativity is free of the **'as if's'** which are at the basis of the corpuscle and electro-magnetic field theories.

That space was curved was a lot to swallow however, and in the absence of a proof that it was or wasn't curved, a question of global significance arose: is Albert Einstein a scientific genius whose stature is comparable to that of Sir Isaac Newton or is he just a crazy old man badly in need of a haircut?

On one hand, Einstein had two major accomplishments in 1905: he had put forth his theory of special relativity; he had successfully explained the photoelectric effect. And further, he had showed in 1915 how his relativistic theories explained the perihelion of mercury.[12] In lieu of these accomplishments the scientific community was not eager to dismiss too easily what Einstein said, no matter how crazy it sounded.

On the other hand, that same community knew full well that Einstein's theories refuted the basic philosophic tenets concerning the nature of space and time and carried with it the responsibility to review these things together with everything dependent on them. It was therefore prone to hold tenaciously to the world views which grew out of classical physics.

It might be said that the irresistible force of Einstein's clarity and logic met the immovable object of physicists' classicism: the ensuing conflict thus germinated attracted worldwide attention.

---

[12] It was known at the time that the orbit of mercury around the sun was not a perfect ellipse as predicted by classical theory but rather a not-quite-perfect ellipse which, in not returning to the same point after each cycle, seemed to precess slowly around the sun. This is known as the perihelion of mercury

... a conflict which intensified for several years smashing tons of chalk into a white powder which was eventually wiped from the forgetful blackboards of the world. In the process however, it became painfully clear that Einstein's relativistic theories were sorely in need of a test.

History had focused a good deal of attention on the issue: the question was, 'is space curved or not?' But it is difficult to answer this question by direct measurement. Luckily, there is the related fact that such a curvature implies that a gravitational field would bend the path of a light beam passing through it. Also a difficult measurement to make but...

... the opportunity finally came four years later in the form of a total eclipse of the sun on May 29, 1919. Another physicist, Eddington, measured the deflection of starlight by the sun's gravity as it passed by the eclipsed region. When the data was analyzed, the results were found to be in keeping with the predictions of general relativity and Einstein became a world renowned celebrity over night.

The premise of general relativity is that the laws of physics must be expressible in a form that applies to *all* coördinate systems in any kind of space.

Einstein concocted the following hypothetical experiment (he had a genius for doing this): suppose a man is in an enclosed elevator sitting on the surface of the earth. Then he would eventually conclude that he is in a gravitational field since every mass m in his system is acted on by a force equal to mg. Now suppose that the elevator is in empty space and that some celestial being is pulling it along giving it an acceleration g. Then every measurement in his space would be the same as it was before and he would still conclude that he was in a gravitational field. The point is that

> *it is impossible to distinguish between a gravitational field and an acceleration field from inside the elevator.*

Einstein concluded that the two were equivalent. The task was to establish the gravitational law in a form which would apply to both systems. This would exhibit the gravitational and acceleration fields as equivalent. In a very deep sense, general relativity makes the single statement, 'gravitational mass and inertial mass are the same'. Their equivalence is already implied by the hypothetical elevator experiment.

To establish the invariance principle, that physical laws be expressed in a form which applies to all coördinate systems, tensor calculus, laden with super- and sub-scripts and many variables is an indispensable tool. The mathematics is intricate and confusing but not conceptually difficult. (Tensors, by the way, are a generalization of vectors.)

The resulting field equations are non-linear and do not lend themselves to an easy solution. Einstein, Infeld and Hoffmann developed a method of successive approximations to the solution of the equations and carried it through to the fourth approximation. It is a feature of this method that each stage ends with a condition which must be satisfied in order that the next approximation exist, i.e. if the method is to produce a solution. It is these conditions which become the equations of physics. It goes like this: the first approximation requires that mass is conserved if the second approximation is to exist; the second approximation requires that $F = ma$ with the F replaced by the universal gravity force if the third approximation is to exist; the third approximation requires that the equations of special relativity hold if the fourth is to exist, etc.

Note that the appearance of the universal gravity force at the end of the second stage constitutes a derivation of the law of gravity. It is no longer a matter of this law being a generalization of experimental data. Additionally, since the law of gravity in this approximation is conjoined to $F = ma$, it might be said that this approximation derives not only the universal gravitational law but also Newton's law of dynamics (at least in part). The conjoining of these two great laws represents a marriage truly made in heaven. The union illustrates the equivalence of gravitational and inertial mass since the same mass plays both roles in the equation.

Finally, we remark that the equations of general relativity are referred to as field equations and the theory itself is referred to as a field theory. It should be emphasized however, that general relativity is not a field theory in the same sense as the Maxwell theory in which a field is created by some conglomerate of masses and/or charges and then a new mass or charge is introduced. It is a field theory in that it establishes the distortion throughout the entire field, (i.e., throughout all space) and this distortion is generated by *all* the masses present.

3.3.2 The Local Gravity Force

3.3.2.1 Motion in the Local Gravity Field

The Local Gravity Constant

It's time to leave the celestial realm now and come back to earth, specifically to its surface where we will study the equations of motion for trajectories in the local gravitational field. Recall from the previous discussion that at the surface of the earth the gravitational field is

$$U \;=\; -G\frac{M}{R^2} \;=\; g$$

where G is the universal gravity constant, M is the mass of the earth, R is the radius of the earth and g is the acceleration due to gravity. To show that g is essentially constant near the surface, consider the differential of U:

$$dU \;=\; 2\,G\,\frac{M}{R^3}\,dR\,.$$

Divide this last equation by the first:

$$\frac{dU}{U} \;=\; -2\,\frac{dR}{R}$$

The last relation illustrates that the fractional change in the force is twice the fractional change in the distance from the center of the earth. Since the radius of the earth is about 4000 miles, the force of gravity changes by about **0.1%** when a body in the gravitational field moves to a height of about twenty miles. In other words, for heights of this magnitude and less above the surface, g is essentially constant.

The approximation is therefore made that the gravity field near the surface of the earth is constant and equal to g. This is called the local gravitational field.

Systems with zero net force are the simplest to consider and have already been discussed. The constant field is next as we proceed up the scale of mathematical difficulty. All constant force fields are essentially identical to one another. The local gravity field is used here as the prototype for all of them.

Constant Acceleration Equations

In preparation for the more general trajectory problem, the equations of motion for a mass moving in one dimension under the influence of a constant force will be considered. Suppose a force of constant magnitude F is applied to a block moving in the positive x-direction at time $t += 0$, and that at that time, the block was at a distance s_0 from the origin and had a velocity v_0. The problem is to describe the motion for positive values of t. *Figure* 3.14 is a diagram showing the initial conditions.

The normal force and the weight cancel each other out in the y-direction, so the net force on the block is $\vec{F}$. Since $\vec{F}$ is constant, the acceleration $\vec{a}$ is constant.

Start with the grammar school equation

$$distance \ = \ rate \ \times \ time$$

For a differential distance ds we have

$$ds \ = \ v \times dt$$

$$\implies \quad v \ = \ \frac{ds}{dt}$$

That makes sense: the velocity is the rate at which the position is changing. Similarly the acceleration is the rate at which the velocity is changing:

$$a \;=\; \frac{dv}{dt}$$

Some constants got lost in the above differential viewpoint. We will now go backwards and pick them up using an integral viewpoint: integrate the last equation

$$v \;=\; \int a\, dt$$

Since the acceleration a is constant it can be taken out of the integral:

$$v \;=\; a \int dt$$

$$v \;=\; at + C$$

where C is the constant of integration.

To find C, note that one of the initial conditions was that when $t = 0, v = v_0$:

$$v_0 \;=\; 0 + C$$

$$\Rightarrow \quad C \;=\; v_0$$

Therefore the equation for the velocity as a function of time (for $t > 0$) is

$$v \;=\; v_0 + at$$

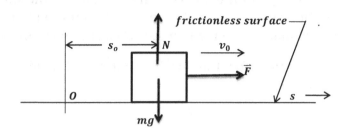

figure 3.14

Equations of Motion for Constant Acceleration

Finally use

$$\frac{ds}{dt} = v$$

to find the position s as a function of time (for $t > 0$):

$$ds = v_0 dt + at\, dt$$

$$\int ds = v_0 \int dt + a \int t\, dt$$

$$s = v_0 t + \tfrac{1}{2}at^2 + C'$$

where C' is the constant of integration. C' can be found from the initial condition that when $t = 0$, $s = s_0$:

$$s_0 = 0 + 0 + C'$$

$$\Rightarrow \quad C' = s_0$$

and the equation for the position as a function of time (for $t > 0$) is

$$. \quad s = s_0 + v_0 t + \frac{1}{2}a\,t^2$$

This last equation expresses the distance in the direction of motion as a function of time: it is the solution to the one-dimensional problem when the acceleration is constant.

The equation for the local gravitational field in three dimensions is

$$\vec{F} \;=\; m\vec{a}$$

$$0\vec{i} \;+\; 0\vec{j} \;-\; mg\vec{k} \;=\; m\!\left(a_x\vec{i} \;+\; a_y\vec{j} \;+\; a_z\vec{k}\right)$$

Recall that two vectors are equal if and only if each of their corresponding components is equal: hence

$$a_x \;=\; a_y \;=\; 0$$

$$a_z \;=\; \text{-}g$$

Therefore the accelerations in the x- and y-directions are both zero: the acceleration in the z-direction is the constant ($\text{-}g$), so the distance traveled as a function of time when the acceleration is constant applies to the z-direction. (The minus sign indicates that the force mg points downward whereas the positive z-direction was chosen upward.) Note that what happens in one direction does not affect what goes on in another. It follows that in Cartesian coördinate Newton's dynamics equation holds in each of the major directions independently of the other two.

Sample Problem: Height of a Bridge over a Ravine

Suppose that the height of a bridge over a ravine is to be determined by dropping a rock from the bridge and measuring the time it takes to hit bottom. If it is dropped at time $t \;=\; 0$ and heard to hit bottom at time $t \;=\; 5$, what is the height of the bridge? What is the speed of the rock when it hits bottom?

Solution: Take $z = 0$ at the bottom of the ravine and the positive direction as up. Then, from the above analysis

$$z(t) = z_0 + v_{0_z}t - \frac{1}{2}gt^2$$

With these definitions, z_0 the initial height is to be determined given that $z = 0$ when $t = 5$. Since the rock was dropped (rather then thrown), the initial velocity was zero:

$$0 = z_0 + 0 - \frac{1}{2}32\, t^2$$

$(g = 32\, ft/sec^2)$ and

$$z_0 = 16 \times 25$$
$$z_0 = 400\, ft.$$

The velocity is found from

$$v_z(t) = v_{0_z} - gt$$
$$v_z(5) = 0 - 32(5)$$
$$v_z(5) = -160\, ft/sec$$

This is the complete solution.

The above results will now be used to verify each of the three basic relations derived earlier from the dynamics equation:

1. *the impulse-momentum relation;*
2. *the work-energy relation;*
3. *the conservation of energy.*

Note that these principles could have been used to solve the problem.)

In the following the subscripts i and f refer to initial and final values.

1.   *Impulse equals change of momentum*

The impulse delivered by the force is (from the definition of impulse)

$$I_z = \int_{t_1}^{t_2} F_z(t)dt$$

$$I_z = -\int_0^5 mg\, dt$$

$$I_z = -m\,32\,(5-0)$$

$$I_z = -160\,m \qquad \text{...............}$$

The change in momentum is

$$\Delta p = p_f - p_i$$

$$\Delta p = m(v_f - v_i)$$

$$\Delta p = -160\,m \qquad \text{..............}$$

$$\Rightarrow \qquad I_z = \Delta p$$

---

2.   *Work equals change in kinetic energy*

The work W done by the force is

$$W = \int_{z_i}^{z_f} F_z\, dz$$

$$W = -mg(0 - z_o)$$

$$W = = -32(0 - 400)\,m$$

$$W = 12,800\,m \qquad \text{.............}$$

The change in kinetic energy is

$$\Delta T = T_f - T_i$$

$$\Delta T = \frac{1}{2} m[(160)^2 - (0)^2]$$

$$\Delta T = 12,800\, m \qquad \text{..............}$$

$$\Rightarrow W = \Delta T$$

---

3. Conservation of energy:

The initial energy H_i is

$$H_i = V_i + T_i$$

$$H_i = mg\, z_i + \frac{1}{2} mv_i^2$$

$$H_i = m\,(32)(400) + 0$$

$$H_i = 12,800\, m \qquad \text{..............}$$

The final energy H_f is

$$H_f = V_f + T_f$$

$$H_f = 0 + \frac{1}{2} m\, v_f^2$$

$$H_f = \frac{1}{2} m\,(160)^2$$

$$H_f = 12,800\, m \qquad \text{..............}$$

$$\Rightarrow H_i = H_f$$

Note that energy is conserved because local gravity is a conservative force field.

It is important to acquire the knack of defining the position of the origin of your coördinate system and the positive direction for your axes and to stick to the definitions you make throughout the problem. Acquaint yourself with the applications of the three principles verified above: it will foster a sense of their respective arenas of application so that, given a problem, you will be able to 'see' a path through one or more of them to a solution.

A second dimension will now be added.

Sample Problem: *Analysis of a Trajectory*

Consider the trajectory of a mass fired from a point P on the surface of the earth in the direction that makes an angle θ with the horizontal (see *figure* 3.15). The task is to find both the horizontal and vertical displacements as functions of time, to determine the trajectory (the shape of the path), and to find the range (the distance it travels before hitting the ground). Find the angle at which the mass should be fired to maximize its range for a given initial thrust.

Solution:

Take the origin at point P where the mass is fired. Let the initial velocity have a magnitude s_0 and let its direction make an angle θ with the positive x-axis.

It was seen before that the motion decomposed naturally into two simpler problems; motion in the z-direction and motion in the x-direction. Consider the z-direction first: the initial displacement is $z_0 = 0$; the initial velocity is $s_0 \sin(\theta)$; therefore the displacement as a function of time in the z-direction is

$$z(t) \;=\; z_0 \;+\; m\, s_0\, sin(\theta)\, t \;-\; \frac{1}{2}\, mg\, t^2$$

(see the previous analysis). Let $t = 0$ when the mass is fired. Hence $z_0 = 0$.

To find x as a function of t, note that the initial speed in the x-direction is $s_0\, cos(\theta)$. There is no force in the x-direction and hence no x-component of acceleration. Since it starts at the origin at $t = 0$, the distance in the x-direction as a function of time is found from 'distance = rate × time':

$$x(t) \;=\; s_0\, cos(\theta)\, t \,.$$

The trajectory is the equation for z in terms of x: we have z in terms of t and x in terms of t. Solve the x equation for t and substitute the result in the z equation. The result will be z in terms of x, the equation of the trajectory:

$$t \;=\; \frac{x}{s_0\, cos(\theta)}$$

$$z(x) \;=\; \left[s_0\, sin(\theta) \;-\; \frac{gx}{2\, s_0 co\, s(\theta)} \right] \frac{x}{s_0 co\, s(\theta)} \,.$$

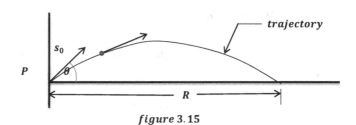

figure 3.15

Trajectory Motion

Thus z is a quadratic function of x, which exhibits the trajectory as an upside down parabolic arc. When the quantity in the brackets equals zero, $z = 0$ and x equals the range, R:

$$R = \frac{2s_0^2 \sin(\theta) \cos(\theta)}{g}$$

$$R = \frac{s_0^2}{g} \sin(2\theta)$$

Since the sine function has a maximum value of $+1$, it is clear that the range has a maximum at the same point:

$$R_{max} = \frac{s_0^2}{g}$$

$$\Rightarrow \quad \sin(2\theta) = 1$$

$$\Rightarrow \quad \theta = 45°$$

The angle for maximum range was determined above by inspection. Formally this value of θ can be determined by a simple application of max-min theory. Consider the range R as a function of θ and note that at the point where it attains its maximum value, the slope of the $(R \, vs \, \theta)$-*curve* is zero. Hence, we have

$$R = R(\theta) = \frac{s_0^2}{g} \sin(2\theta)$$

To find the value of θ that maximizes R, apply the condition

$$\frac{dR}{d\theta} = 0$$

$$\frac{2 s_0^2}{g} \cos(2\theta) = 0$$

$$\Rightarrow \quad \theta = 45°$$

Another quantity of interest is the maximum height h attained. One could conjecture (correctly) that this is the height at half the total flight time but the answer obtained would be a conjectured one unless there was more justification. Another way is to use the conservation of energy. (The local gravity force is conservative with a potential energy $V = mgz$.) If the potential energy is to be zero at ground level, then the initial energy (the energy at firing time) is all kinetic. Use the subscript f to indicate the highest point of the trajectory. At this point the z-component of velocity is zero and the x-component (constant throughout the motion) is $s_0 cos(\theta)$. Therefore, the energy equation is:

$$H_i = H_f$$

$$T_i + V_i = T_f + V_f$$

$$\frac{1}{2} ms_0^2 + 0 = \frac{1}{2} ms_0^2 cos^2(\theta) + mgh$$

$$\Rightarrow h = \frac{s_0^2 sin^2(\theta)}{2 g}.$$

Trajectory problems are usually concerned with:

1. *the decomposition of the motion into two simpler ones, one in the z-direction and one in the x-direction;*
2. *the use of the displacement, velocity and acceleration formulas to find x and z as functions of time;*
3. *relating the time equations to the trajectory equation;*
4. *the recognition that energy is conserved and the use of the energy relations whenever possible.*

There are a number of variations on trajectory problems, but they are all analyzed in essentially the same way as shown here.

3.3.2.2   Constrained Motion in the Local Gravitational Field

Problems of this kind involve inclined planes, pulleys, strings and the like which place geometric constraints on the system. Usually surfaces are frictionless so energy is conserved.

For this type of problem, two things need to be developed:

1. *the ability to draw a good force diagram representative of the problem;*
2. *the habit of gathering enough information directly from the physical principles and constraint conditions so that the number of equations equals the number of unknowns.*

It is expedient to do these things first to avoid confusing the physics and the algebra. In some situations, the acceleration turns out to be constant. When this is the case, the equations for the displacement and velocity discussed above apply. But the acceleration need not be constant, for example, when a block moves over a curved surface. In these cases, it is usually the conservation of energy that enables you to find the quantities of interest.

Recall that if gravity is the only force that does work, (normal forces do no work) then energy is conserved. The equation for the conservation of energy is a good ace-in-the-hole to get you out of tight situations.

A survey of the following sample problems will illustrate the methods by which these problems are solved.

Sample Problem 1: Block Sliding down an Inclined Plane

A block slides down an inclined plane which makes an angle θ with the horizontal, then across a frictionless table top to a wall. Upon striking the wall, the block sticks to it. What is the impulse delivered to the wall by the block?

Analysis:

(Refer to *figure* 3.16.) It is hopeless to attempt to find the details of the force $f(t)$ exerted by the wall on the block: hence, the impulse I cannot be found from its integral definition. However, it is known that the impulse (the time-integrated version of $f(t)$) is the change in momentum. The final momentum is zero, so the problem involves finding the momentum just prior to impact. This would be known if the speed were known. The speed can be determined from the initial height because energy is conserved prior to impact.

Hence, find the speed at table-top level from the energy equation; find the momentum of the block just prior to impact; find the impulse from the change in momentum.

Before doing the calculations, note the forces shown in the diagram. Since the surfaces are frictionless, the contact force is normal to the surface and hence, does not affect the speed of the block. The only force that does work is the conservative gravity force. (While the block is on the inclined plane, the normal force conspires with that component of the gravity force which is normal to the plane, to keep the block on the surface.)

185

Solution:

Let the z-axis be perpendicular to the table top and the positive x-axis along the table top to the right. Let h be the vertical distance through which the CM of the block moves from its starting position to its level along the table. Then

$$H_i \;=\; H_f$$

$$T_i + V_i \;=\; T_f + V_f$$

$$0 + mgh \;=\; \frac{1}{2}\,m\,v_x^2 + 0$$

$$\Rightarrow \quad v_x \;=\; \sqrt{2gh}$$

Now, let subscripts i and f refer to just prior to and just after the block hits the wall. The momentum prior to impact is

$$p_i \;=\; m\,v_x$$

$$p_i \;=\; m\,\sqrt{2gh}\;.$$

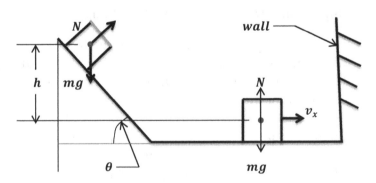

figure 3.16

Block on a Frictionless Surface

After impact, the momentum is zero. The impulse delivered to the block is found from the impulse-momentum relation:

$$I \;=\; \Delta p$$

$$I \;=\; p_f \;-\; p_i$$

$$I \;=\; 0 \;-\; m\sqrt{2gh},$$

$$I \;=\; -m\sqrt{2gh}.$$

The impulse received by the block was in the negative *x*-direction which is reasonable since it stopped the motion of the block. Keep in mind that we are talking about the motion of the block and therefore, the impulse ***received*** by the block. The action-reaction law says that the wall received an equal and opposite impulse delivered by the block. Some care must be exercised here to avoid a mistake in sign.

Note that the angle of incline was not needed to solve the problem. Also the block could have slid down a curved surface through the same vertical distance and the result would have been the same. "How can this be?" you say. If you know, then we are in pretty good shape in this regard but if you think you don't, I confront this judgment, claiming that you actually do, but you just haven't connected with your own knowledge yet. Recall from ***section*** 1.3.2, that potential energy was definable only when the work done by the associated force was independent of the path of motion. This is one of those cases: the kinetic energy when the block reached the table top was calculated using the potential energy ***mgh***, of the (conservative) local gravitational field and the potential energy change associated with this change in the vertical height is independent of the path the block took to move from height h to height 0. So, inclined plane or frictionless roller-coaster track, falling through a vertical distance h results in the same speed when the block reaches the table top.

Exercise:

Discuss what happens when the block goes across the corner at the join of the inclined plane and the table top. Is the above analysis affected? Explain.

Sample Problem 2: Two Masses Attached by a String over a Pulley

A block of mass m_1 is on a frictionless inclined plane which makes an angle θ with the horizontal. It is attached to a second mass m_2 by a massless string over a massless pulley as shown in *figure* 3.17. Describe the motion of the system.

Analysis:

Isolate m_1 and let T_1 be the tension in the string where it is attached. The normal force N cancels with that component of m_1g which is normal to the incline of the plane. So consider the components parallel to the plane choosing the positive direction as upwards along the incline. For motion in this direction:

$$F = ma$$

$$T_1 - m_1g\,sin(\theta) = m_1a_1 \qquad (1)$$

There are two unknowns, T_1 and a_1.)

Isolate m_2 and let T_2 be the tension in the string where it is attached. Choose positive direction downwards (this is not necessary) to correspond to the positive up direction along the plane for the first block. For downward motion of the second mass:

$$m_2 g \; - \; T_2 \;\; = \;\; m_2 a_2 \qquad\qquad (2)$$

There are now two equations and four unknowns, T_1, T_2, a_1 and a_2.)

We must have

$$T_1 \;\; = \;\; T_2 \qquad\qquad (3)$$

because the massless string and frictionless pulley cannot support a difference in tension anywhere along the string. (This equation and the next one are equations of constraint.)

(There are now three equations in four unknowns.)

For the last equation, we use the condition of constraint: the string is of constant length, and hence

$$a_1 \;\; = \;\; a_2 \qquad\qquad (4)$$

There are now 4 equations in four unknowns. The physics is done. The remaining task is to solve the 4 simultaneous algebraic equations.

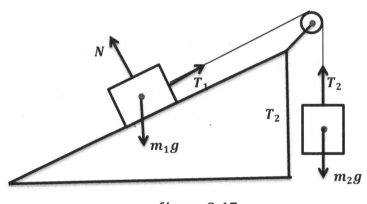

figure 3.17

Constrained Motion of Two Blocks

Solution:

Simplify the set to two equations in two unknowns right from the start by letting the two tensions equal T and the two accelerations equal a. (This uses the information in **eq's (3)** and **(4.)**) Then

$$T \; - \; m_1 g \sin(\theta) \;\; = \;\; m_1 a \qquad\qquad (1')$$

$$m_2 g \; - \; T \;\; = \;\; m_2 a \qquad\qquad (2')$$

T can be eliminated by adding **(1')** and **(2')**. Then solve for a:

$$a \;\; = \;\; \frac{m_2 - m_1 sin(\theta)}{m_1 + m_2} \; g$$

This describes the motion: the system moves with constant acceleration displayed here as a fraction of g. It moves in the positive direction (i.e., m_1 moves up the plane) if $m_2 > m_1 sin(\theta)$; it has acceleration 0 if

$m_2 = m_1 sin(\theta)$; it moves in the negative direction (i.e., m_1 moves down the plane) if $m_2 < m_1 sin(\theta)$.

Discussion:

Using the last value of a in either of the two previous expression, the tension T is found to be

$$T = \frac{m_1 m_2 [1 + sin(\theta)]}{m_1 + m_2} g$$

It sometimes helps to define dimensionless parameters to express your results. These are pure numbers. For example m_1 and m_2 both have the dimensions of a mass; their ratio however is a pure number.

Let

$$\mu = \frac{m_2}{m_1}.$$

Then the acceleration a, and the tension T, written in terms of μ are

$$a = \frac{\mu - sin(\theta)}{\mu + 1} g;$$

$$T = \frac{1 + sin(\theta)}{\mu + 1} m_2 g.$$

The fractions on the right of each of the last two expressions are now pure numbers. Hence, from the first equation it is seen that a and g must have the same units: from the second it is seen that T and $m_2 g$ must have the same units This is a slightly more ordered way to express your results.

Also, it is sometimes good to check your results partially by considering extreme conditions which can often be assessed by inspection to see if they 'make sense'. We proceed to do that now by supposing that the angle of the incline is **90°**. This distortion of the geometry produces a situation in which the two masses are simply strung over a pulley and the inclined plane has no effect. In this extreme

$$sin(\theta) \quad = \quad 1$$

$$a \quad = \quad \frac{\mu - 1}{\mu + 1} g$$

$$T \quad = \quad \frac{2}{\mu + 1} m_2 g \,.$$

The equation for the acceleration indicates that the motion is in the positive direction when the second mass is larger and that the acceleration gets smaller as the second mass gets closer to the first. The acceleration is zero when the masses are equal. Also interchanging the masses reverses the sign of the direction of motion. It is seen further that when the masses are equal, $(\mu = 1)$ the tension in the string is $m_2 g$ which is what is required to balance the gravity force at each of the masses. Considerations such as these afford some insights into the problem and serve as a partial check on your analysis.

Sample Problem 3: Mass Attached to a String Pinned to the Ceiling

A mass m_1 is attached to a string of length **R** pivoted at a point **P**. With the string taut, the mass is dropped from a point level with the pivot. When it is directly below the pivot, it strikes a mass m_2 initially at rest (see *figure* 3.18). Assuming the collision to be perfectly elastic, find the maximum angle θ with the vertical to which the mass m_1 rises after the collision. Evaluate your answer when applied to each the three special cases: $m_2 =$

0; $m_2 = m_1$, $m_2 = \infty$; . Discuss each to ascertain that your answer makes sense.

Analysis:

The first mass falls under the influence of the local gravity force. Since the tension in the string is always perpendicular to the direction of motion, it does no work. It follows that energy is conserved between the initial condition and the time just prior to impact. During this period, only the gravity force enters into the energy equation. At impact, momentum is conserved from the instant before to the instant after. Since the collision is perfectly elastic, energy is also conserved. These two conditions at impact afford sufficient information to calculate the velocity of each mass immediately after collision. Therefore, the kinetic energy of m_1 after impact can be determined and the mass will rise until all this energy is potential. This is the condition which determines the maximum angle to which it rises.

This problem can be solved using only the conservation laws of energy and momentum.

Solution: Taking the height to be zero at impact level, application of the energy conservation equation from the initial state to impact produces the speed just prior to impact:

$$T_i + V_i = T_f + V_f$$

$$0 + m_1 g R = \frac{1}{2} m_1 v_1^2 + 0$$

$$\Rightarrow \quad v_1 = \sqrt{2gR}$$

Letting primes denote the quantities after collision, the equations obtained by applying the conservation of momentum and the conservation of energy to the collision (just before to just after) are:

conservation of momentum

$$\dot{P}_i \;=\; P_f$$

x component

$$m_1 v_1 \;=\; m_1 v_1' \;+\; m_2 v_2' \qquad (1)$$

....

Conservation of energy

$$H_i \;=\; H_f$$

$$m_1 v_1^2 \;=\; m_1 v_1'^2 \;+\; m_2 v_2'^2$$

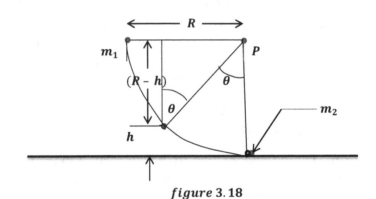

figure 3.18

Local Gravity, Collisions and Energy Transfers

Let μ equal the mass ratio m_2/m_1, then (1) and (2) become

$$v_1 \;=\; v_1' \;+\; \mu v_2' \qquad (1')$$
$$v_1^2 \;=\; v_1'^2 \;+\; \mu v_2'^2 \qquad (2')$$

Exercise: Solve the last two equations for v_1' and v_2' to obtain:

$$v_1' = \frac{1-\mu}{1+\mu} v_1$$

$$v_2' = \frac{2}{1+\mu} v_1$$

The next step is to obtain an expression for the angle $\boldsymbol{\theta_{max}}$, the maximum angle to which the first mass rises after collision. The conservation of energy equation will yield the result: take the initial state as that right after the collision and the final state as that when the mass is at its maximum height, **h.**

$$H_i = H_f$$

$$T_i + V_i = T_f + V_f$$

$$\tfrac{1}{2} m_1 v_1'^2 + 0 = 0 + m_1 gh$$

We have the previous results that:

$$v_1' = \frac{1-\mu}{1+\mu} v_1 ;$$

$$\Rightarrow v_1 = \sqrt{2gh}$$

Using these results in the equation that precedes them

$$\tfrac{1}{2} m_1 \left[\tfrac{1-\mu}{1+\mu}\right]^2 2gR = m_1 gh$$

$$\Rightarrow \qquad \frac{h}{R} = \left[\frac{1-\mu}{1+\mu}\right]^2$$

From the diagram we have

$$\cos(\theta_{max}) = \frac{R-h}{R}$$

$$\cos(\theta_{max}) = 1 - \frac{h}{R}$$

$$\cos(\theta_{max}) = 1 - \left[\frac{1-\mu}{1+\mu}\right]^2$$

$$\cos(\theta_{max}) = \frac{4\mu}{(1+\mu)^2}$$

This last equation determines the angle θ_{max}. Since μ is always positive, the cosine is always positive. Hence,

$$-\frac{\pi}{2} \leq \theta_{max} \leq +\frac{\pi}{2}$$

We now check the three special cases:

case 1: $(m_2 = 0), (\mu = 0)$

There is no mass for m_1 to collide with. It just keeps going past the point where the collision was supposed to occur without changing its speed. It would also be expected that it return to the same height from which it started, only on the other side. These expectations are supported by the equations:

$$v_1' = \frac{1-\mu}{1+\mu} v_1$$

$$\Rightarrow \quad v_1' = v_1 \quad \text{when} \quad \mu = 0$$

$$\cos(\theta_{max}) = \frac{4\mu}{(1 + \mu)^2}$$

$$\Rightarrow \theta_{max} = 90° \quad \text{when } \mu = 0$$

case 2: $(m_2 = m_1), (\mu = 1)$

This is the straight on collision of two identical masses. m_2 acquires the speed that m_1 had and m_1 stops. The maximum angle is therefore zero. The equations below bear this out:

$$v_1' = \frac{1-\mu}{1+\mu} v_1$$

$$\Rightarrow v_1' = 0 \ when \ \mu = 1$$

$$\cos(\theta_{max}) = \frac{4\mu}{(1+\mu)^2}$$

$$\Rightarrow \theta_{max} = 0^o \ when \ \mu = 1$$

Case 3: $(m_2 = \infty), (\mu = \infty)$

m_2 is a locomotive engine; m_1 is a ping pong ball. m_1 just bounces back and rises to its initial position at $\theta_{max} = -90°$. m_2 does not move. The equations support these claims:

$$v_1' = \frac{1-\mu}{1+\mu} v_1$$

$$\Rightarrow v_1' \to -v_1 \ when \ \mu \to \infty$$

$$v_2' = \frac{2}{1+\mu} v_1$$

$$\Rightarrow v_2' \to 0 \ when \ \mu \to \infty$$

$$\cos(\theta_{max}) = \frac{4\mu}{(1+\mu)^2}$$

$$\Rightarrow \theta_{max} \to -90° \ when \ \mu \to \infty$$

Exercise: Carry the previous problem through when the collision is perfectly inelastic, i.e., when the masses stick together upon impact. Show that

$$\cos(\theta_{max}) = 1 - \frac{1}{(1+\mu)^2}$$

Take the analysis as far as you can: discuss the special cases as was done in the previous problem making sure that your equations 'make sense'; find the energy lost on impact and check that your equation makes sense in the limiting cases. (The limiting cases are, $m_2 = 0$, $m_2 = m_1$ and $m_2 = \infty$, or in terms of the mass ratio $\mu = m_2/m_1$, $\mu = 0$, $\mu = 1$ and $\mu = \infty$.

Sample Problem 4: Equations of Motion for Two Masses Attached by a String over a Pulley

A mass m_1 is attached to another mass m_2 by a string over a massless pulley. The masses are initially at rest side by side at a height Z above a table top. At $t = 0$, the system is allowed to move; when m_2 hits the table, m_1 continues to rise. What is the maximum height attained by m_1? At what time is it attained? Assume $m_2 > m_1$. Express your answer in terms of the mass ratio $\mu = \dfrac{m_1}{m_2}$.

Analysis:

Take $z = 0$ at the table top and measure the height from there (refer to *figure* 3.19). This problem will make use of the displacement, velocity and acceleration formulas for a constant force field. (It will be seen that the system moves with constan t acceleration.) Let $t = 0$ when the system is released; let $t = t_1$ when the second mass hits the table; let t_2 be the duration from the time the second mass reaches the table to the time the first mass reaches its maximum height.

Set up the equations of motion from a force diagram and use them the find the acceleration of the system. The time t_1 when the second mass hits the table can be found from the displacement equation when the acceleration is constant. Find the velocity at this time and use it as the initial velocity for the second part of the motion, from table-top crash to maximum height of m_1. Find t when the speed is zero. This is the time t_2, the duration from the table top crash to time the first mass attains maximum height. Then the time t_0 for the first mass to attain its maximum height is $(t_1 + t_2)$.

Solution:

Apply Newton's dynamics equation to each of the asses to find the acceleration of the system:

$$F = ma$$

$for\ m_1$: $T - m_1 g = m_1 a$ (1)

$for\ m_2$: $m_2 g - T = m_2 a$ (2)

(Note: some eliminating has already been done on this set of equations. There are actually two tensions to be considered and two accelerations, one of each for each mass. The two tensions were set equal because in this situation, the string will not support a difference in tension anywhere along its length; the two accelerations were also set equal because they are constrained to be so by the string which cannot cha nge its length. (See **sample problem 2.**)

We are left with two equations in two unknowns, T and a. T is eliminated by adding them together. Then the resulting equation can be solved for the acceleration. In terms of the mass ratio $\mu = m_1/m_2$, the result is

$$a = \frac{1 - \mu}{1 + \mu} g$$

Exercise: Fill in the missing algebra to obtain this last result.

The last relation shows that the acceleration of the system is constant: thus the displacement and velocity of the system are determined by the constant acceleration equations. (See the *section 3.3.2.1*, '**Equations of Motion When the Acceleration is Constant**') Everything will be left in terms of the acceleration a, until the end. At that time the above expression will be used.

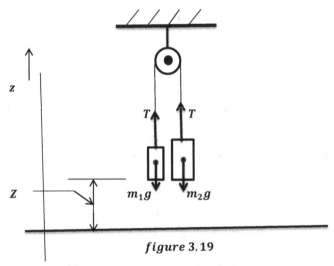

figure 3.19

Two Masses Connected Over a Pulley

Use the displacement formula for constant acceleration to

$$z \quad = \quad z_0 \ + \ v_0 t \ + \ \frac{1}{2} \, a t^2$$

$$Z \quad = \quad 0 \ + \ 0 \ + \ \frac{1}{2} \, a t_1^2$$

The last equation determines the time t_1 it takes to travel a distance Z when the system starts at rest and the acceleration a is constant. The duration from $t = 0$ to the time the mass m_2 reaches the table is

$$t_1 = \sqrt{\frac{2Z}{a}}.$$

The velocity, v_1 at that time is

$$v_1 = v_0 + at_1$$

$$v_1 = 0 + \sqrt{2aZ}$$

Since gravity is the only force acting after impact the velocity during the second part of the motion is given by

$$v(t) = v_1 - gt$$

The velocity when it reaches its maximum height is zero, so t_2 must satisfy the equation

$$0 = \sqrt{2aZ} - gt_2$$

$$t_2 = \frac{\sqrt{2aZ}}{g}$$

The time t_0 to reach its maximum height is

$$t_0 = t_1 + t_2$$

$$t_0 = \sqrt{\frac{2Z}{a}} + \frac{\sqrt{2a}}{g}$$

The expression for the acceleration a found earlier was

$$a = \frac{1-\mu}{1+\mu} g$$

If this expression is used in the above, the result after some algebis

$$t_0 = 2 \sqrt{\frac{1 + \mu^2}{1 - \mu^2}} \sqrt{\frac{Z}{g}}$$

Finally, we find the maximum height attained by m_1 by applying the displacement formula for a constant acceleration. At impact, the height is $2Z$. Therefore, taking $t = 0$ at that time,

$$z(t) = z_0 + v_1 t - \frac{1}{2} g t^2$$

$$z_{max} = 2Z + v_1 t_2 - \frac{1}{2} g t_2^2$$

$$z_{max} = 2Z + \frac{2aZ}{g} - \frac{aZ}{g}$$

$$z_{max} = \left[2 + \frac{a}{g}\right] Z$$

After substituting the expression for a and simplifying, the final result is

$$z_{max} = \left[\frac{3 + \mu}{1 + \mu}\right] Z$$

That was a lot of detailed fussing around to come to this answer. We might ask, "Is there an easier way?" The answer is 'yes'.

Note that everywhere in this analysis only the local gravity force was in play and consequently energy was conserved - -- *except* for the collision of m_2 with the table. Therefore, if we knew how much energy was lost (i.e., no longer available to the motion) in this collision, we could account for it in the energy equation and solve the problem essentially in one step. Letting H represent the energy, we have

$$H_i - H_{lost} = H_f$$

If we take $z = 0$ at table top level, then at the instant of impact, m_2 had no potential energy and it had a kinetic energy of $\left(\frac{1}{2} m_2 v_1^2\right)$. This is the only energy lost to the system throughout the motion. To obtain the result an easier way, begin as above, until you find v_1 and apply the above equation.

At that point in the analysis, we had two partial results:

$$a = \frac{1-\mu}{1+\mu} g \; ;$$

$$v_1 = \sqrt{2aZ} \,.$$

To apply the above equation, note that the initial and final energies were entirely potential (of the mgz-type). Hence

$$H_i - H_{lost} = H_f$$

$$m_2 gZ - \frac{1}{2} m_2 v_1^2 = m_1 g z_{max}$$

or, dividing by m_2

$$\mu gZ + gZ - \frac{1}{2} v_1^2 = \mu g z_{max}$$

Using the values for v_1 and a found earlier,

$$(1 + \mu)gZ - \frac{1}{2} 2 \frac{(1-\mu)}{(1+\mu)} gZ = \mu g z_{max}$$

$$\left[(1+\mu) - \frac{1-\mu}{1+\mu}\right] Z = \mu z_{max}$$

$$\Rightarrow \quad z_{max} = \left[\frac{3+\mu}{1+\mu}\right] Z \,,$$

which agrees with our previous result.

Exercise: Fill in all the missing algebra steps in the above.

Thought Problem: [13]

The diagram of *figure* **3.20** shows two tracks. The first is horizontal and a cart C_1 is free to move along it; the second starts horizontal then dips as shown and returns to its original level. A second cart C_2 identical to the first is free to move along this track. Assume all motions frictionless. If both carts are given an initial velocity v at $x = 0$ and both stay on their respective tracks, determine whether C_1 and C_2 get to $x = L$ at the same time or C_1 gets there first or C_2 gets there first. Explain your answer fully. Make sure your explanation **proves** that your choice is correct

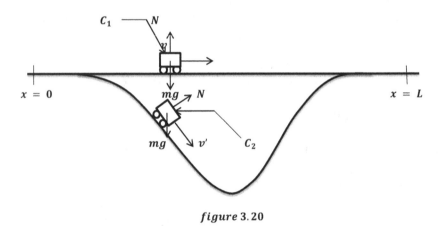

figure 3.20

The Cart Race --- Place your bets!

---

[13] This problem first appeared as part of the oral exam for a doctorate degree at M.I.T. Its purpose was to test the candidate's knowledge of the basic laws of mechanics. I apologize that I do not know the author's name. He certainly deserves recognition.

3.4   SIMPLE HARMONIC MOTION

3.4.1   Restoring Forces

3.4.1.1   Linear Restoring Forces and Stable Equilibrium

In *section* 2.3 there was an extensive discussion of equilibrium. Small displacements from equilibrium were considered and the requirement for the equilibrium to be stable was that the displacements give rise to forces that tended to bring the system back to its equilibrium state. Forces such as these are called *restoring forces.*

Consider a displacement from a state of equilibrium and let it be sufficiently small so that only first order differentials of the distances need be considered. Since the equilibrium is stable it is evident that the forces act in the opposite direction from the displacements: only then will the forces tend to return the system to equilibrium. Hence, the restoring forces for the situation described are negatively proportional to the displacements. Since they are linear functions (that's what 'proportional' means) of the displacements, they are called *linear restoring forces.* We have,

> *a linear restoring force is a restorative force negatively proportional to the displacement of a system from equilibrium.*

Incidentally, the remarks above indicate that it is more than an unfounded conjecture and less than a proof that

> *a sufficiently small displacement of any system from a state of stable equilibrium subjects the system to restoring forces that are linear.*

Chapter III: Applications to Linear Motion

This statement is actually a good and most useful rule of thumb for those who are faced with the problem of analyzing the stability of complex systems. **Perturbation theory** is the field addressed to this investigation. Perturbations (small disturbances from equilibrium) are often used, first, to determine whether or not a system is stable, and if it is, to study the system's recovery from the occurrence of factors that upset its balance. The second italicized statement above indicates that for sufficiently small displacements, the restoring forces are linear and the analysis can be based on the simple results to be obtained below.

The subject matter is the motion of mechanical systems under the action of linear restoring forces. The study is germane to all wave phenomena, oscillatory motions, resonance, perturbations, etc. and applies to everything from sub-atomic particles to galaxies.

A spring is the simplest mechanical device which exerts a linear restoring force when stretched or compressed (within its elastic limits) from equilibrium length. Consequently it will be the central object of investigation in this section. It will serve to introduce the basic concepts of wave motions and oscillations as well as the parameters which are used to describe them. The same parameters and concepts pervade all fields of application. So while we are focused on springs, know that the subject matter has applications far beyond the particular confines observed here.

3.4.1.2 Spring Forces and Potential Energy

Attention is now focused on forces of the form

$$F = -kx,$$

where x is the displacement from equilibrium and the restoring force is proportional to it. The proportionality constant k is called the spring constant. The minus sign indicates that the force opposes the displacement: if the displacement is positive, the force is in the negative direction and visa versa. Hence, the force tends to restore the equilibrium state which corresponds to a zero displacement ($x = 0$): these are the defining characteristics of a *linear restoring force*.

Since the motions under consideration are one-dimensional, i.e., they refer to displacements in a single direction and since the force depends only on position, the work done in going from a displacement x_1 to another displacement x_2 is independent of the path you traverse. You could, for example go halfway and rest a while or overshoot your mark and come back or any configuration of back-and-forth's and start-and-stops that start you out at x_1 and end you up at x_2: the total work done is always the same, depending only on the endpoints, x_1 and x_2. The same thing said in mathematical terms is, "The work integral in no way references the path of motion:

$$W = \int_{x_1}^{x_2} \vec{F} * d\vec{s}$$

The very fact that the work can be computed without any reference to the path of motion means that it cannot depend on it. (There are such things as time-varying potentials but we will not be concerned with those here. For us, if the work integral does not depend on the path of motion, it depends only on the endpoints of the motion.) In the case at hand the spring force is always in the direction of motion and the work integral is

$$W \;=\; -k \int x\,dx$$

$$W \;=\; -\frac{1}{2}\,kx^2$$

The negative of the function on the right is called the potential energy $V(x)$. So the work integral is already evaluated once and for all. In this and similar cases, it is said that the force is **'*derivable from a potential*'** or that **'*the force is conservative*'**.

More correctly, we have

$$V(x) \;=\; \frac{1}{2}\,kx^2 \;+\; C$$

The constant of integration, C can be chosen arbitrarily since only **differences** in potential energy are of physical significance. In taking any such difference, the constant cancels out:

$$V(x_2) \;-\; V(x_1) \;=\; \frac{1}{2}\,kx_2^2 \;+\; C \;-\; \frac{1}{2}kx_1^2 \;-\; C$$

$$V(x_2) \;-\; V(x_1) \;=\; \frac{1}{2}\,kx_2^2 \;-\; \frac{1}{2}kx_1^2 \,.$$

The simplest choice is $C = 0$. This sets the zero point of potential energy at the equilibrium length, $x = 0$. (See *sections* $1.3.1 - 1.3.3$.)

Recall that since the potential energy is the integral of the force in some sense, the force must be the derivative of the potential energy is some sense. This is nothing more than the reversal of the operation of finding the potential energy from the force. It was found that the force was what is called the directional derivative of the potential energy.

(It is in that particular sense that the force is 'derived from a potential'.) If the direction of motion is the x-direction, then

$$F_x = -\frac{\partial V(x)}{\partial x}$$

$$F_x = -\frac{d}{dx}\left[\frac{1}{2}kx^2\right]$$

$$F_x = -kx$$

The directional derivative of **section 2.3.2** reduces to the ordinary derivative here because there is only one direction to consider.

As already indicated, the spring is to be used as the prototype for this kind of force. Consider what happens on an atomic level when a spring is elongated or compressed. Each molecule is initially in an equilibrium position with respect to the conglomerate of molecules in its vicinity. The compression or elongation of the spring has produced a slight perturbation of the state by pushing the molecules closer together or pulling them farther apart than they would like to be. The molecules then exert forces on one another producing a tendency for the material to return to its initial state. The net effect (i.e., the sum of all these forces) on a macroscopic scale is a restoring force proportional to the displacement from equilibrium.

By no means are the intermolecular forces of the material linear. The linearity of the macroscopic effect derives from the fact that the individual molecules are only *differentially* displaced: the macroscopic displacement is the sum of these. What we have here is an example of a complex system differentially perturbed from its equilibrium. The displacements are sufficiently small so that only first order differential effects are in play: these are the linear ones. The spring constant k is the macroscopic manifestation of these forces and hence is dependent on the shape of the spring and the material from which it is made.

This indicates that there is a real limit on the operation of the spring action of any material: namely, that it is possible to exceed the point where the displacements can be considered 'small'. Subjecting a spring to extravagant elongations or compressions will do exactly this and permanently deform the molecular arrangement of the material. This is referred to as *'exceeding the elastic limit of the spring'*.

The spring force $F = -kx$ is therefore accurate only within a certain range of x values. We will consider only situations in which the spring does not exceed its elastic limit. The only exceptions to this rule will be those cases where some resonance effect makes the system fly apart.

3.4.1.3 The Motion of a Mass Attached to a Spring

Consider the diagram of $figure\ 3.21$. A mass m is shown attached to a spring whose other end is secure. When the mass is displaced a distance x from its equilibrium position at $x = 0$ and then released, it is free to move on a horizontal frictionless surface in one direction (the x-direction). The only net force acting on the mass therefore is that from the displacement of the spring from its equilibrium.

The Differential Equation of Motion

The equation for the motion of the mass is

$$F \quad = \quad ma$$

$$-kx \quad = \quad m\frac{d^2x}{dt^2}$$

or, in its more usual form,

$$\frac{d^2x}{dt^2} + \frac{k}{m}x \quad = \quad 0$$

We want to know how the position $x(t)$ varies with time given that the only information we have from the dynamics equation is that x must satisfy the above relation between itself and its second time derivative. An equation of this sort is called a differential equation. If you have not yet had a course in differential equations, the prospects of finding a solution may look pretty slim. But don't judge too hastily; take another look.

Developing a Solution

The quantity k/m is a constant and we will worry about that shortly. The essential feature of the equation is that if it has a solution $x(t)$, then $x(t)$ plus its second derivative add to zero. Now if you have had a course in elementary differential calculus, you already know of two such functions: $sin(t)$ and $cos(t)$. Each of these satisfies the condition that its second derivative is its negative. So even without a course in differential equations, there is considerable hope of finding a solution to this particular one.

This makes sense also from the standpoint that we are dealing with springs (restoring forces in general) and would therefore expect things to go back and forth.

Let's pause here and see what we've got. Work with just one of the functions, $x(t) = sin(t)$, say, by introducing it into the equation to see if it works, i.e. to see if it is a solution. The result is:

$$\frac{d^2x}{dt^2} + \frac{k}{m}x = 0:$$

Substitute $sin(t)$ for $x(t)$:

$$-sin(t) + \frac{k}{m}sin(t) = 0.$$

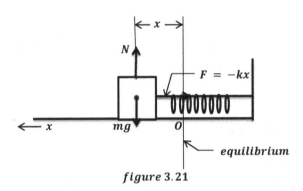

figure 3.21

Motion of a Spring-Mass System

This is **not** a solution. Note that there are values of t which satisfy this equation but that is not sufficient for $x(t) = sin(t)$ to qualify as a solution. To see this, note that the above differential equation is just $F = ma$ in the disguise of a mass-spring system and $F = ma$ must hold instant by instant. That tells us that to get a solution it is necessary to find a function $x(t)$ which when substituted into the equation is **identically** zero so that it agrees with the right side at **every** time **t**.

We didn't do badly though. Our first stab at the solution has the correct functional dependence: $sin(t)$ and $-sin(t)$ add to zero at any time **t**. They almost cancelled to produce a function which was identically zero: the only glitch was the constant k/m.

We now attempt glitch-elimination: consider a slight variation of our first attempt, namely, let $x(t) = sin(\omega t)$ where ω is a constant. (Maybe we can pick a value for the constant ω which will help out.) Then we have:

$$\frac{d^2x}{dt^2} + \frac{k}{m}x = 0.$$

Substitute $sin(\omega t)$ for $x(t)$:

$$-\omega^2 sin(\omega t) + \frac{k}{m}sin(\omega t) = 0.$$

We can get a solution from this, but clearly, not just any old ω will do: the left side of the above equation is identically zero only if

$$\omega = \sqrt{\frac{k}{m}}.$$

It follows that a solution that works is

$$x(t) = sin\sqrt{\frac{k}{m}}\,t.$$

A more general solution is obtained if it is noticed that if $x(t)$ satisfies the differential equation then so does $Ax(t)$ where A is an arbitrary constant. To see this, suppose some function $f(t)$ satisfies **eq(1)**; evaluate the left side of

the differential equation inserting the function
$g(t) = A f(t)$:

$$\frac{d^2 A f(t)}{dt^2} + \frac{k}{m} A f(t) =$$

$$A \frac{d^2 f(t)}{dt^2} + \frac{k}{m} A f(t) =$$

$$A \left[\frac{d^2 f(t)}{dt^2} + \frac{k}{m} f(t) \right]$$

Since $f(t)$ satisfies the equation, the bracketed expression on the last line is zero. Hence the first line is also zero which illustrates that $Af(t)$ also satisfies the equation.

We have essentially accomplished what we set out to do: a solution to the differential equation has been found. (It will be elaborated on below.)

Notice that the constant ω tells us how fast the mass will move back and forth. It is an important parameter of oscillatory motion: ω is called the **angular velocity** of the motion and is measured in radians/second. From the equation we see that given the spring constant k and the mass m, the angular speed (radians per second) of the oscillatory motion is already determined.

The value of ω found above is the **natural angular velocity** of the spring-mass system and will be designated ω_o to distinguish it from other angular frequencies which may be in play.

Another important parameter of oscillatory motion is the **period:**

> *the period t of an oscillatory motion is the time it takes to go through one complete cycle.*

The condition to determine the period T is that when $t = T$, the angle in the sine function must be 2π:

$$\sqrt{\frac{k}{m}} \, T \; = \; 2\pi$$

$$\Rightarrow \quad T \; = \; 2\pi \sqrt{\frac{m}{k}}$$

The **frequency** is also an important parameter. The **frequency** is the number of cycles per second. A little thought reveals that the frequency is the reciprocal of the period. For example, if the period is **1/10** of a second, then there are **10** cycles per second. If f represents the frequency, then

$$f \; = \; \frac{1}{T}.$$

The last parameter of interest is the arbitrary constant A. The sine function varies from -1 *to* $+1$: if it is multiplied by a constant A, the resulting function is a sine wave that varies from *–A* to *+A*. A is called the **amplitude** of the wave.

To summarize what precedes, it is the angular velocity ω_o that comes most naturally from the differential equation of motion: it is expressed in radians/second, and is proportional to the frequency which is expressed in cycles/second. Since one cycle is 2π radians, $\omega_o = 2\pi f_o$[14] The period which is the time for one complete cycle is expressed in seconds/cycle and is the reciprocal of the frequency (cycles/second).

---

[14] The particular frequency f_o that corresponds to ω_o is called the natural frequency of the system. It is the frequency at which the system oscillates in the absence of outside influences.

The three quantities ω_0, f_0 and T are redundant. Only one of them is independent: i.e. if one is known then the other two can be determined from it. Remember that ω_0 the angular velocity is the most basic of the three since it relates directly to the physical quantities k and m through the dynamics equation. The following two relations contain the information that must be remembered:

$$\omega_0 \;\; = \;\; \sqrt{\frac{k}{m}}$$

$$\omega_0 \;\; = \;\; 2\pi f \;\; = \;\; \frac{2\pi}{T}$$

Note that the solution of the differential equation that was obtained is always zero when $t = 0$. This seems a little restrictive since it should be possible to define the zero point in time arbitrarily; not just at one of those times when the mass happens to be at zero displacement.

Finding the Most General Solution

What went wrong? Did $F = ma$ sell us short? The answer is 'no'. The problem is that while we obtained *a solution* of the differential equation, we did not find *the most general solution* and consequently it will not fit the most general situation that may give rise to the equation in the first place. Recall that we started with the solution $x(t) = sin(\omega_0 t)$ and found that if $x(t)$ was a solution then $Ax(t)$ was also a solution for any arbitrary constant A. The class of functions that satisfy the equation can be further enlarged: note that the functions $sin(\omega_0 t)$ and $cos(\omega_0 t)$ are identical except for the fact that one is displaced from the other by an amount $\pi/2$ along the t-axis. Hence if $sin(\omega_0 t)$ is a solution so is $cos(\omega_0 t)$ and, as before, so is $B cos(\omega_o t)$ where B is an arbitrary constant. The most general solution that can be formed thus far is

$$x(t) \quad = \quad A\, sin(\omega_o t) \; + \; B co\, s(\omega_o t)$$

Exercise: Verify that the last expression satisfies the differential equation (1), where A and B are arbitrary constants and $\omega_o = \sqrt{k/m}$.

The exercise justifies the use of the above expression as a solution of the differential equation that governs the motion of the spring-mass system. Note that the constants A and B are left arbitrary by the solution of the dynamics equation. This is exactly as it should be: they are the degrees of freedom necessary to make the solution fit any given set of initial conditions.

Solving for the Constants of Integration

(Fitting the Arbitrary Constants to the Initial Conditions of the Problem)

For example, if $t = 0$ when the mass is at its equilibrium position traveling in the negative x-direction with a speed s, then the constants A and B are made to fit these conditions as follows.

Start with the solution above:

$$x(t) \quad = \quad A\, sin(\omega_o t) \; + \; B co\, s(\omega_o t)$$

Using the first condition, $x(t) = 0$ when $t = 0$:

$$0 \quad = \quad 0 + B \qquad \Longrightarrow \qquad B = 0$$

What remains is:

$$x(t) \quad = \quad A\, sin(\omega_o t)$$

The second condition is $v(t) = -s$ when $t = 0$:

$$v(t) = \frac{dx(t)}{dt}$$

$$v(t) = A\omega_o \cos(\omega_o t)$$

$$v(0) = A\omega_o$$

$$-s = A\omega_o$$

$$\Longrightarrow \quad A = -\frac{s}{\omega_o}$$

The constants A and B that make the solution fit the initial conditions of this specific problem have now been found. Therefore, the solution is

$$x(t) = -\frac{s}{\omega_o} \sin(\omega_o t)$$

where $\omega_o = \sqrt{k/m}$.

It is generally the case that all but the simplest problems in mechanics involve finding the solution of differential equations and these solutions invariably produce arbitrary constants whose values are assigned to fit the initial (or some set of) conditions which characterize the particular situation.

The material in this section can be summarized as follows.

It was found that the application of $F = ma$ to a mass moving under the influence of a spring requires that the motion $x(t)$ of the mass satisfy the differential equation

$$\frac{d^2 x}{dt^2} + \omega_0^2 x = 0$$

where

$$\omega_0 = \sqrt{\frac{k}{m}}$$

The most general solution to this equation is

$$x(t) = A\cos(\omega_0 t) + B\sin(\omega_0 t)$$

where **A** and **B** are arbitrary constants *(these are the constants of integration).* They are determined in each case to agree with the particular conditions of the problem.

The natural oscillation rate of the system is determined by the spring constant **k** and the mass **m**.

3.4.1.4 The Parameters of Oscillatory Motion

Whenever we speak of oscillatory motion or wave motion which is closely related, words and expressions like *amplitude, frequency, phase, angular velocity* and *period* are used. These terms are defined and described in this section.

PERIOD: The period of an oscillation is the duration of time it takes to go through one complete cycle:

$$T = 2\pi\sqrt{\frac{m}{k}}$$

FREQUENCY: The frequency is the number of cycles/second; it is the reciprocal of the period which is the number of seconds/cycle:

$$f = \frac{1}{T} = \frac{1}{2\pi}\sqrt{\frac{k}{m}}$$

ANGULAR VELOCITY: The angular velocity is proportional to the frequency. The frequency is cycles/second; the angular velocity is radians/second:

$$\omega_0 \;\; = \;\; 2\pi f \;\; = \;\; \sqrt{\frac{k}{m}}$$

AMPLITRUDE: the amplitude of an oscillation is the value of its maximum displacement from equilibrium. Since both the sine and cosine functions have amplitude **1**, the amplitude of

$$A\, sin(\omega_0 t) \quad or \quad A\, cos(\omega_0 t)$$

is **A**. The amplitude of

$$A\, cos(\omega_0 t) \;\; + \;\; B\, sin(\omega_0 t)$$

is

$$\sqrt{A^2 \; + \; B^2}$$

This last assertion is discussed in the following discussion of phase angle.

Phase Angle

The *phase angle* requires more explanation. It refers to the same solution we have been talking about but in a different form. We have to do some trig now.

Recall from trig that the sine and cosine curves are exactly the same shape; they are simply displaced from one another by **90°:**

$$cos(s - 90°) \;\; = \;\; sin(s)$$

One might say that a sine wave is just a cosine wave with a phase angle of **90°**. This is what a phase angle is --- a displacement in the angle of a sine or cosine function. It need not be constant in general but for our purposes it is:

> *a phase angle is a constant angular displacement in the argument of a sine or cosine function.*

The curve of $cos(s - \varphi)$ is the same as the curve $cos(s)$ except that the former is shifted by an angle φ in the positive s-direction.

Attention is now focused on a phase-shifted cosine function of amplitude C; using the trig formula for the cosine of the sum of two angles,

$$C\cos(\omega_0 t - \varphi) = C\cos(\varphi)\cos(\omega_0 t) + C\sin(\varphi)\sin((\omega_0 t)$$

$$C\cos(\omega_0 t - \varphi) = A\cos(\omega_0 t) + B\sin(\omega_0 t)$$

where the substitutions

$$A = C\cos(\varphi)$$
$$B = C\sin(\varphi)$$

have been made. An important property of sines and cosines has just been illustrated:

> *the sum of two sinusoids having the same frequency is another sinusoid of the same frequency. The effect of the addition is to change the phase and alter the amplitude.*

It is necessary to acquire the ability to go from one form of the answer to the other. The substitutions above suggest the simple diagram shown in *figure* 3.22. if C and φ are given then A and B can be found from the substitution equations above; if A and B are given then C and φ can be found from

$$C = \sqrt{A^2 + B^2}$$

$$tan(\varphi) \;=\; \frac{B}{A}$$

$A \;=\; Ccos(\varphi)$ $\qquad\qquad$ $C \;=\; \sqrt{A^2 + B^2}$

$B \;=\; Csin(\varphi)$ $\qquad\qquad$ $tan(\varphi) \;=\; \frac{B}{A}$

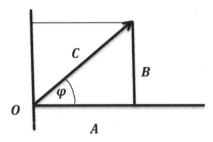

figure 3.22

Parameter Relations: (A,B) $\longleftrightarrow$ *(C,φ)*

Where did the diagram come from? The equations for **A** and **B** indicate that a vector of length **C** in the **xy**-plane making an angle **φ** with the positive **x**-axis has **x**- and **y**-components equal to **A** and **B** respectively. The four above equations used to transform from the (**A, B**)-form to the (**C, φ**)-form or visa versa can be read off the diagram. It's easier to remember the diagram than to memorize the equations. The resulting equation for **C** is the relation cited before without proof.

The two forms for expressing **x(t)** are equivalent in every way. Occasion to exploit the (**C, φ**)-form will present itself in the derivation of the equations that describe the phenomenon of resonance.

3.4.1.5   Sample Problems Involving Springs

The solution $x(t)$ of the differential equation of motion constitutes a complete solution of the mass-spring problem. Knowing $x(t)$, you can find the velocity $v(t)$ and the acceleration $a(t)$ by differentiation. Many of the problems involving springs are therefore addressed to the question, 'do you know how to interpret and use this solution?'

To analyze spring problems, remember these four things:

1. *you know $x(t)$ from the start so you can find any quantity concerning the motion;*
2. *you must be able to find the undetermined constants to fit any set of conditions specific to the current problem;*
3. *you must know the definitions of the parameters and the relations between them;*
4. *he spring force is conservative with potential energy $\frac{1}{2}k x^2$, so use the energy conservation equation whenever possible.*

The following sample problems are presented to illustrate some of the details of analysis.

Sample Problem 1: Collision between a Moving Mass and a Spring

A mass m moves along a frictionless surface with velocity v_o in the positive x-direction. It comes into contact with a spring of spring constant k attached to a wall as seen in *figure* 3.23. What is the maximum compression of the spring? What is the time from impact to maximum compression? Describe the motion.

Analysis:

Once the mass and spring are in contact, the spring compresses according to the equation for $x(t)$ derived earlier. The maximum compression is the amplitude of the motion: it can be determined by taking $t = 0$ at the instant of contact and using the initial conditions that at $t = 0$, the compression is zero and the velocity is v. The time to compress can be determined in any one of three ways: find the time t when $x(t)$ is maximum; find t when $v(t) = 0$; find $t = T/4$.

A second method for solving the problem which is somewhat easier is to use the energy equation. It will determine the maximum compression in one step. The time from impact to maximum compression is $T/4$ which can be determined directly from the mass m and the spring constant k.

The description of the motion is as follows. The mass moves inertially until it reaches the spring. Then the spring compresses slowing the mass until its velocity is zero. At this instant all the energy of the system is stored in the spring as potential energy. The force delivered by the compressed spring acts on the mass pushing it in the direction from which it came until the mass leaves the influence of the spring. All the potential energy is returned to the mass and is now kinetic. Since the spring force is conservative, the final kinetic energy and hence the speed of the mass have the same values as initially.

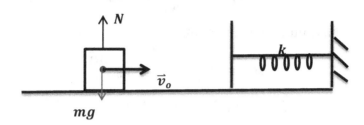

figure 3.23

Dynamic Spring Compression

Solution 1: Using the Equation of Motion

$$x(t) = A\cos(\omega_0 t) + B\sin(\omega_0 t)$$

At $t = 0$ (the instant of contact) the displacement is zero; the speed is v:

$$x(0) = A + 0$$
$$0 = A + 0$$
$$\Rightarrow A = 0.$$

Hence

$$x(t) = B\sin(\omega_0 t)$$
$$v(t) = \frac{dx(t)}{dt} = B\omega_0\cos(\omega_0 t)$$
$$v_0 = v(0) = B\omega_0$$
$$v_0 = B\omega_0$$
$$\Rightarrow B = \frac{v_0}{\omega_0}$$

The constants have now been determined so that the solution fits the conditions of the problem. The displacement as a function of time is

$$x(t) \;=\; \frac{v_0}{\omega_0}\sin(\omega_0 t)$$

The maximum displacement is

$$B \;=\; \frac{v_0}{\omega_0} \;=\; v_0\sqrt{\frac{m}{k}}$$

The time t' of maximum compression is

$$t' \;=\; \frac{T}{4}$$

$$t' \;=\; \frac{2\pi}{4}\sqrt{\frac{m}{k}}$$

$$t' \;=\; \frac{\pi}{2}\sqrt{\frac{m}{k}}.$$

Solution 2: Using the Energy Equation

Let the subscript i refer to the instant of contact; the subscript f, to the time of maximum compression:

$$H_f \;=\; H_i$$

$$\frac{1}{2}\,k\,B^2 \;=\; \frac{1}{2}mv_0^2$$

$$\Rightarrow\; B \;=\; v_0\sqrt{\frac{m}{k}}$$

Maximum compression occurs at

$$t' \;=\; \frac{T}{4}$$

$$t' \;=\; \frac{2\pi}{4\omega_0}$$

$$\Rightarrow \quad t' \;=\; \frac{\pi}{2}\sqrt{\frac{m}{k}}$$

Note that (as usual) it is easier to use the energy relations rather than the equations of motion.

Now consider the following situation. The assertion concerning it is the subject of the next problem. Let a given spring have equilibrium length L and spring constant k. Consider now that a mass m is suspended from the spring in the local gravitational field and the system is in equilibrium for some increased length L' for which the spring force exactly balances the gravity force.

It is a property of this system that the restoring force when the mass is displaced a distance x from its **new equilibrium** position is still $-kx$. Hence, when the spring force and the gravity force are acting in the same direction, all the previous results concerning the equation of motion still apply

>*if the gravity force is ignored and displacements are measured from the new equilibrium point.*

This is the result to be established in ...

Sample Problem 2: Mass-Spring System in a Gravity Field

A mass m is suspended from the ceiling by a spring of spring constant k. Find the difference ΔL between its normal equilibrium length L and the new equilibrium length L'. Show that if the mass is displaced a distance x from its new equilibrium length (a distance L' from the ceiling) that the restoring force is still $F = -kx$.

Analysis:

To find the new equilibrium length, balance the spring force and the weight; to find the restoring force displace the mass a distance *x* from its new equilibrium length and find the net force.

Note: This problem is simple but the result is important.

Solution: (refer to *figure* 3.24.)

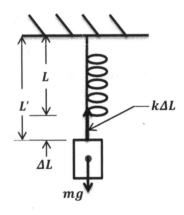

figure 3.24

Gravity and Spring Dynamics

$$k\Delta L \;=\; mg$$

$$\Rightarrow \quad \Delta L \;=\; \frac{mg}{k}$$

This answers the first question. Now let the origin be at the new equilibrium and let x be measured downward. Displace the mass a distance x downward so the net force is the gravity force minus the spring force:

$$F = mg - k(x + \Delta L)$$

$$F = mg - kx - k\frac{mg}{k}$$

$$F = -kx$$

Since x is the displacement from t new equilibrium, this last expression implies that all the preceding material about mass-spring systems can be applied

> *if displacements are measured from the new equilibrium length and the gravity force is ignored.*

This result applies to spring-mass systems in any constant force field.

Sample Problem 3: Collision between a Falling Mass and a Vertical Spring

A spring whose spring constant is k is on the floor in a vertical position and a mass m is dropped from a height h above the top of the spring. What is the maximum compression M of the spring in the ensuing motion? Assume the spring-cup assembly massless. (Refer to *figure* 3.25.)

Analysis:

This problem will be solved in two ways: the first using the equations of motion; the second, using the energy equation.

Method 1: Using the Equations of Motion

The spring and gravity forces act in the same direction. Therefore the equation of motion $x(t)$ applies only if displacements are measured from the new equilibrium point and the gravity force is ignored. This is the situation from the instant the mass reaches the spring until maximum compression.

Method 2: Using the Energy Equation

Since the energy equation does not reference the motion of the spring it can be applied directly without the artifice of measuring displacements from the new equilibrium point and ignoring the gravity force. Since energy is conserved throughout, the energy before the mass is dropped is the same as the energy at maximum compression.

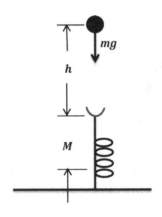

figure 3.25

Weight Dropped on a Vertical Spring

(**Prediction of things to come:** the second solution will be easier.)

Solution 1.

Find the velocity at the instant of contact from the energy equation. Start when the ball is dropped: end at the instant of contact:

$$H_i = H_f$$

$$T_i + V_i = T_f + V_f$$

$$0 + mgh = \frac{1}{2}mv^2 + 0$$

$$\Rightarrow v = \sqrt{2gh}$$

Once the ball makes contact, the displacement as a function of time is

$$x(t) = A\cos(\omega_0 t) + B\sin(\omega_0 t).$$

If the positive direction is defined as **up** and $t = 0$ at the instant of contact, then A and B are determined as follows.

$$x(0) = A.$$

But (see **sample problem 2**)

$$x(0) = \frac{mg}{k}$$

$$\Rightarrow A = \frac{mg}{k}$$

$$v(t) = -A\omega_0 \sin(\omega_0 t) + B\omega_0 \cos(\omega_0 t)$$

$$v(0) = B\omega_0$$

From the above, $v(0) = -\sqrt{2gh}$

$$-\sqrt{2gh} = B\omega_0$$

$$B = -\frac{\sqrt{2gh}}{\omega_0}$$

231

$$\Rightarrow \quad B \quad = \quad -\sqrt{\frac{2mgh}{k}}$$

The amplitude of the motion is

$$C \quad = \quad \sqrt{A^2 + B^2}$$

$$C \quad = \quad \sqrt{\left(\frac{mg}{k}\right)^2 + \frac{2mgh}{k}}$$

The maximum compression M of the spring is

$$M \quad = \quad \frac{mg}{k} + C$$

$$\Rightarrow \quad M \quad = \quad \frac{mg}{k}\left[1 + \sqrt{1 + \frac{2hk}{mg}}\right].$$

Solution 2:

Let M be the maximum compression of the spring. and let the potential energy (mgh) due to gravity be zero at the point of maximum compression. Then:

$$H_i \quad = \quad H_f$$

$$T_i + V_i \quad = \quad T_f + V_f$$

$$0 + mg(h + M) \quad = \quad 0 + \frac{1}{2}kM^2$$

This is a quadratic equation in M with solution

$$\Rightarrow \quad M \quad = \quad \frac{mg}{k}\left[1 + \sqrt{1 + \frac{2hk}{mg}}\right].$$

Note that the use of the energy equation not only saves a considerable amount of work but also affords a conceptually simpler approach to the problem.

232

3.4.1.6   The Phenomenon of Resonance

Until now we have been talking about what is referred to as *free vibrations* or *free oscillations*. These are the responses of the system allowed to react on it own after being given some initial conditions which perturb its equilibrium state. It does so in its own characteristic way with its own natural frequency. Now we are going to investigate a different kind of excitation: instead of imparting some set of initial conditions to the system, the system is going to be continually subject to an outside influence. This influence is referred to as *forced oscillations*. It will be sinusoidal in nature.

The reaction of the system depends on the excitation frequency ω: when $\omega \ll \omega_0$, the system follows the excitation drive as if the spring were not there; in the other extreme, when $\omega \gg \omega_0$ the system is 180° out of phase with the excitation drive always moving in the direction opposite to the external force; when ω is almost equal to the natural frequency, $\omega \approx \omega_0 = \sqrt{\dfrac{k}{m}}$, a curious phenomenon known as resonance occurs. At resonance, the system reacts wildly, limited to finite oscillations only by the friction force. For systems having sufficiently small friction each cycle of the excitation imparts more energy to the system and the system literally explodes. Hence, the singer can shatter the glass by singing the musical tone that matches the resonant frequency of the glass and holding it long enough for the energy imparted to the glass to become critical; and hence, a wind whose force oscillates at the natural frequency of a bridge can cause it to flap wildly like a piece of cloth in a breeze and fly apart. The critical point is reached when the excitation energy imparted to the object in question makes the object vibrate with an amplitude which exceeds its elastic limit.

A detailed analysis of resonance is presented below. It will be seen that the spring-mass system is a proto-typical representative of the phenomenon as it exhibits all the characteristic features. The approach is to solve $\vec{F} = m\vec{a}$ for the spring-mass system determining the response for a fixed but unspecified excitation frequency. The resulting equations that describe the response will show its dependence on the particular drive frequency chosen. Then it will be possible to examine the response of the system for different drive frequencies.

A friction force proportional to the velocity will be included in the analysis because when the drive frequency is near the natural frequency of the system, the friction force affects the response of the system in non-negligible ways. It will be seen that the response of the system is such that both the amplitude and phase of its motion depend on the excitation frequency.

The system is set up as shown in *figure* **3. 26**. The mass **m** sits on a surface and is attached to a spring; the other side of the spring is attached to a wall which is driven back and forth sinusoidally. The surface is such that the mass is subject to a friction force proportional to its velocity, always in the direction opposite to it.

When the system is started up, it will 'try' to oscillate at its natural frequency thus fighting the excitation whose frequency is different. This is the result of the 'starting up' process. The resulting motion which is complex and difficult to track is referred to as the ***transient response*** and is not of interest here. The system will eventually settle down into a more organized and repetitive movement called the **steady state response.** In the steady state, the system has given up its natural tendencies and having no choice, oscillates at the excitation frequency.

The Differential Equation of Motion:

Let $x = 0$ correspond to the equilibrium position of the mass when the wall is at its central position at $y = 0$. The wall will be moved back and forth sinusoidally with amplitude C and angular velocity ω. Consider the system at some arbitrary time t when the displacement of the mass is x and the displacement of the wall is y.

Then the force acting on the mass from the spring is

$$F = -k(x - y)$$
$$F = -k(x - C\cos(\omega t))$$

The friction force is

$$fr = -\mu v(t)$$
$$fr = -\mu \frac{dx}{dt}$$

where μ is a proportionality constant. The dynamics equation then is

$$F = ma$$

$$-kx + kC\cos(\omega t) - \mu \frac{dx}{dt} = m \frac{d^2x}{dt^2}$$

Divide the equation by m and use the relation $\omega_0{}^2 = \frac{k}{m}$.

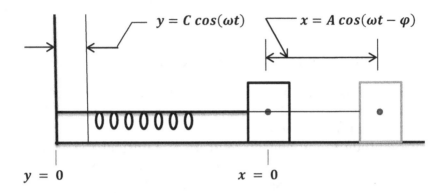

figure 3.26

Response of a Spring-Mass System to Forced Vibrations

After rearranging the terms,

$$\frac{d^2x}{dt^2} + \frac{\mu}{m}\frac{dx}{dt} + \omega_0^2\, x \;=\; \omega_0^2\, C\,cos(\omega t)$$

(Don't get upset when I say we are going to guess the solution to this equation. It's not as hard as it looks. Besides when you take your first course in differential equations, you will find that most of them are solved by organizing them into groups each one of which has a certain form for its solution which ultimately was guessed at in much the same as we are doing here. A course in differential equations might very well be called, 'how to guess the solution in nine easy lessons and one hard one'.)

The Solution:

Ready?...Set?...Look at the above equation and note that essentially (ignoring the constants for now) we have *x* and its first and second derivatives added together and the sum is a *cos(ωt)* term.

We already know that all the derivatives of *cos(ωt)* produce only *sin(ωt)* functions and *cos(ωt)* functions. We might therefore expect that the unknown function *x(t)* is a sinusoid of the same frequency (angular velocity *ω*) as the drive. Hence, we are prompted to try a solution of the form

$$x(t) \;=\; A\cos(\omega t - \varphi).$$

That is the most general sinusoid having the same frequency as the drive. If there is no solution of this form, it will not work when we insert it into the differential equation. If, on the other hand, it satisfies the equation, then our educated guess will have paid off and we will have a solution. (Secretly, I already know it's going to work, so stick with it.)

Inserting the above assumed form for the solution into the differential equation to see if it works is *almost* the plan. Notice that the three terms on the left will be sines and cosines of the sum of two angles. That will lead to a trigonometric mess. We are going to eliminate two-thirds of that mess by the following.

The above solution *x(t)* assumes that the wall had zero phase angle and that the block was out of phase by an angle *φ*. There is no loss of generality if we assume that the block has no phase angle and that the wall is out of phase with it by an angle (−*φ*). Then there will be only one cosine-of-the-sum-of-two-angles term. This amounts to choosing $t = 0$ at a different time and does not affect the results.

We therefore consider the situation equivalent to the original in which the basic $F = ma$ equation is replaced by

$$\frac{d^2x}{dt^2} + \frac{\mu}{m}\frac{dx}{dt} + \omega_0^2 x = \omega_0^2 C \cos(\omega t + \varphi)$$

with the guessed-at solution

$$x(t) = A \cos(\omega t).$$

Note that when an arbitrary phase angle is introduced, it doesn't make any difference whether $x(t)$ is a sine function or a cosine function: each choice will put its own conditions on the phase angle in such a way as to make the final results identical.

All the preliminaries are done now. To continue, substitute $x(t)$ into the equation and expand the right side using the formula for the cosine of the sum of two angles:

$$-A\omega^2 \cos(\omega t) - A\frac{\mu}{m}\omega \sin(\omega t) + A\omega_0^2 \cos(\omega t) =$$

$$\omega_0^2 C \cos(\varphi) \cos(\omega t) - \omega_0^2 C \sin(\varphi) \sin(\omega t)$$

Collect all the $sin(\omega t)$ terms into one term and all the $cos(\omega t)$ terms into another:

$$\left[A(\omega_0^2 - \omega^2) - \omega_0^2 C \cos(\varphi)\right] \cos(\omega t) +$$

$$\left[-A\frac{\mu}{m}\omega + \omega_0^2 C \sin(\varphi)\right] \sin(\omega t) = 0$$

Now the crucial step: the left side of this equation must be identically zero (i.e. for all values of t). This is equivalent to saying that $F = ma$ must hold at every instant during the motion. But notice that the last equation is of the form

$$X \cos(\omega t) + Y \sin(\omega t) = 0$$

and it has already been determined that the form on the left equals a sinusoid of the same frequency with amplitude

$$C = \sqrt{X^2 + Y^2}$$

Since the right side of the equation is 0, C must equal zero. But $C = 0$ if and only if *each* of the quantities X and Y is zero. We therefore have the two conditions:

$X = 0$:

$$A(\omega_0^2 - \omega^2) - \omega_0^2 C \cos(\varphi) = 0 \qquad (1)$$

$Y = 0$:

$$A\frac{\mu}{m}\omega - \omega_0^2 C \sin(\varphi) = 0 \qquad (2)$$

We have two equations in two unknowns: A, the amplitude of the block's response; and φ, the phase angle between the drive and the block. They are solved as follows.

The Amplitude and Phase Responses:

To find A, the amplitude of the response, transpose the second term of each of the above equations, square the equations and add them:

$$A^2\left[(\omega_0^2 - \omega^2)^2 + \left(\frac{\mu\omega}{m}\right)^2\right] = \omega_0^4 C^2$$

(The identity, $sin^2(\theta) + cos^2(\theta) = 1$, was used.).

Solve for A:

Amplitude Response:

$$A = \frac{\omega_0^2\, C}{\sqrt{\left(\omega_0^2 - \omega^2\right)^2 + \left(\frac{\mu\omega}{m}\right)^2}}\;;$$

to find φ, the phase angle, divide (**2**) by (**1**):

Phase Response:

$$tan(\varphi) = \frac{\mu\omega}{m(\omega_0^2 - \omega^2)}.$$

As will be seen shortly, these two results, the amplitude and phase of the block's response to the forced oscillations contain a great deal of information about the phenomenon of resonance.

These equations indicate that the amplitude of the response is different for each drive frequency and the same is true for the phase angle between the drive and the response. The amplitude and phase as functions of the drive frequency are shown in the sketches of $figure\,3.27$. (Refer to both the sketches and the equations during the following discussion.) The coefficient of friction μ is generally a small number and will be neglected when compared to the larger quantities in the expressions. It becomes important when the drive frequency ω is near the natural frequency $\omega_0 = \sqrt{\frac{k}{m}}$, of the system. It is in this vicinity that resonance occurs and it is here that the damping force alone keeps the amplitude finite.

We will now examine the results in three cases of interest:

1. *the drive frequency is very small:* $\omega \ll \omega_0$
2. *the drive frequency is very large:* $\omega \gg \omega_0$

3. *the drive frequency is near resonance:* $\omega = \omega_0$

Case 1: this is the case when the movement is very slow` compared to the reaction of the spring-mass system. It would be expected that the movement of the mass simply follow the movement of the drive i.e., the whole system moves as a unit. That this is so can be seen from the following expressions which approximate the amplitude and phase relations for small values of the drive frequency ω. They are

$$A = \frac{C}{1 - \left(\frac{\omega}{\omega_0}\right)^2}$$

$$tan(\varphi) = \frac{\mu\omega}{k\left(1 - \left(\frac{\omega}{\omega_0}\right)^2\right)}$$

The first of these relations exhibits A as nearly equal to C the amplitude of the drive: as $\omega \to 0$, $A \to C$ *and* $tan(\varphi) \to 0$. As the drive frequency increases but remains small, both the amplitude and phase increase.

Case 2: this is the case when the drive is very fast so that the spring-mass system can hardly respond before the drive is going the other way. It would be expected that the amplitude of the response get smaller and eventually go to **0** as the drive frequency increases and that the drive and response be **180°** out of phase. For very large drive frequencies, the expressions for the amplitude and phase are approximated by:

$$A = \left(\frac{\omega_0}{\omega}\right)^2 C$$

$$\tan(\varphi) = \frac{-\mu}{m\omega}$$

As $\omega \to \infty, A \to 0$ *and* $\varphi \to 180°$. This agrees with the judgments made above.

Case 3: the drive frequency is near the natural frequency of the system. This is the condition for resonance. It would be expected that the response amplitude become very large. The amplitude and phase in this case are approximated by:

$$A = \frac{m\omega_0 C}{\mu}$$

$$\tan(\varphi) = \infty$$

$$\Rightarrow \quad \varphi = 90°$$

The only thing that keeps the amplitude finite is the damping coefficient μ. If μ is sufficiently small, the amplitude will be large enough to exceed the elastic limit of the spring and the system will explode. The phase relation indicates that the drive and response are **90°** out of phase when the drive frequency ω is near the natural frequency ω_0 of the system.

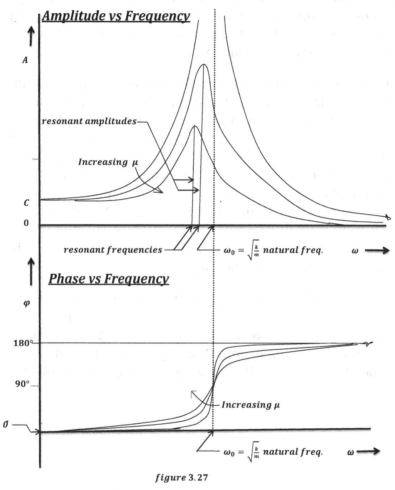

figure 3.27

Amplitude and Phase Responses to Forced Vibrations

The first of the two graphs above shows three curves: each is a plot of the amplitude response vs. drive frequency for a given value of the damping coefficient *μ*. For each of these curves, the drive frequency at which the response is maximized is called the **resonant frequency**: the maximum response at this frequency is called the **resonant amplitude.** In the absence of sufficient damping, it is the resonant amplitude which may cause the system to exceed its elastic limit and explode.

It can be seen from these curves that as the damping increases, the resonant amplitudes decrease in magnitude and the frequency at which resonance occurs deceases, moving further away from the system's natural frequency. This indicates that with increased damping, the system is less likely to undergo an explosion due to resonance. On the other hand as $\mu \rightarrow 0$, the resonant amplitude $A_r \rightarrow \infty$ and the resonant frequency $\omega_r \rightarrow \omega_0$. It follows that for sufficiently small damping, explosive resonance is certain to occur at some drive frequency in the vicinity of the system's natural frequency ω_0.

The amplitude and phase graphs of *figure* 3.27 contain all the information essential to the understanding of the phenomenon of resonance. They should be studied until this understanding is imparted to you. If you decline, your wine glass will shatter and your bridge will collapse.

CHAPTER IV

The Angular Parameters of Motion

4.1 ORIENTATION

4.1.1 Kinematics

In *section* 1.2.4 it was determined that when a force is acting on a moving body, its two components, one tangential to the direction of motion and one normal to it have separate and mutually exclusive effects on the motion: the tangential component changes only the speed of the body; the normal component changes only its direction. It might be asked, 'what is the nature of this decomposition by which the complex problem of describing the general motion is broken down into two simpler problems?'

To answer this question, suppose for a moment that we had no knowledge whatsoever of the dynamics equation $\vec{F} = m\vec{a}$ and we were interested only in the kinematics, i.e. specifically in the geometric relation between the velocity and acceleration.

Let $\vec{S}$ be any vector in space and add to it a differential change $\Delta\vec{S}$. The vector change $\Delta\vec{S}$ can be in any direction but can always be resolved into two components: $\Delta\vec{S}_t$ in the direction of the original vector; and $\Delta\vec{S}_n$ in a direction normal to the direction of the original vector. Consider each of these components by itself. It is obvious from the geometry that $\Delta \vec{S}_t$ affects a change in the length of $\vec{S}$ without altering its direction; $\Delta \vec{S}_n$ affects a change in the direction of $\vec{S}$ without changing its length (considering only first order differentials). This is the property we are talking about.

The last statement is true for any vector in two or more dimensions. In particular, it is true for any velocity vector. Note that the velocity vector always points in the direction of the line tangent to the path of motion. The differential *change* in the velocity which can be in any direction can be resolved into components in the tangential and normal directions the first of which is in the direction of the velocity. These correspond to the two components of the acceleration vector $\vec{a}_t$ and $\vec{a}_n$. If this is considered together with the remarks of the previous paragraph it is seen that the property we are talking about descends from the kinematic relation between differential velocity changes and accelerations.

Now consider the dynamics equation which exhibits the force and acceleration vectors as proportional. The pOroportionality indicates that these two vectors are parallel. By equating the corresponding components, it becomes evident that the tangential and normal components of force inherit the property of which we are speaking from the acceleration: thus the component of force $\vec{F}_t$ in the direction of motion changes only the speed of the moving body and the component of force $\vec{F}_{n,}$ normal to the path of motion changes only the direction of motion.

These remarks trace the lineage of speed and direction changes from velocity vector changes to accelerations through the kinematics of motion, and from the acceleration components to the causal force components through Newton's dynamics equation.

4.1.2 The Choice of Parameters

The task of determining the effect of the tangential component of force on the speed resulted in the work-energy relation which tells us quantitatively exactly how the speed is related to this component. The entire analyses of work, kinetic energy and potential energy are based on the consideration of the tangential component of this natural decomposition.

The task at hand is to consider the effects of the other component, the normal one. The effect of this component of acceleration will be analyzed by considering the constant speed circular motion of a point mass. The fact that the speed is constant implies that there is no tangential component of acceleration (force); the fact that the motion is circular implies that the normal (radial) component has a constant non-zero value.

It would seem that a coherent approach to the analysis of many mechanical problems could proceed comfortably on the basis of this natural decomposition into tangential and normal components but this unfortunately is not the case: the general description of the motion of a body in three dimensions in terms of tangential and normal components of acceleration is encumbered by the problems of space geometry and leads to differential equations which are difficult to solve. A description of the problems encountered by such an approach is given in *section* **4. 8.**

This is not to imply however, that the tangential-normal viewpoint is useless. The fact is that the principles derived from its analysis in the discussion of work and kinetic energy remain in faithful servitude to the solution of mechanics problems regardless of the particular way in which forces are resolved into components. In those instances when it is not feasible to resolve forces into tangential and normal components, some other set of orthogonal directions must be chosen.

Keep in mind that one of the objectives here is the application of $\vec{F} = m\vec{a}$ to the rotational motion of a solid body. Imagine then, a solid rotating around an axis that passes through its *CM.* Each of its points exhibits circular motion traversing a distance and having a speed and acceleration all determined by its distance from the axis of rotation. It is a problem in kinematics to describe the motion of each of these points.

It should be noted in this regard that it is possible to take advantage of the fact that the body is a solid: the angle θ through which the body rotates is a single parameter which is common to each of its points, i.e. each point of the body exhibits an angular displacement θ in its own particular circle of motion. This follows from the fact that the relative positions of the internal points of a solid have an unchanging relation to each other. This of course is not true for a liquid or gas.

For this reason, θ is used as a descriptor to represent the rotational motion of a solid. Any particular point in the solid is related to the linear distance *s* traversed along the circumference of its circle of motion by the equation

$$s \;=\; r\theta,$$

where r is its distance from the axis of rotation.

Recall that there is considerable reason to decompose a general motion into the motion of the *CM* plus another motion around the *CM*. The latter component of this decomposition uses the angular parameters of motion which are defined by the above relation and its time-derivatives.

This is certainly not the only set of parameters which specify the position and orientation of a body: any number of other parameter sets could be devised. But recall that the analysis of the motion of the *CM* revealed that it moves according to Newton's laws. This fact strongly suggests that the position of the *CM* be among the parameters used to describe the motion; given this choice, the axis through the *CM* around which the body is rotating is almost demanded as a next choice; and finally, given both of these choices, the angular displacement around this axis is the single parameter that completes the specification of the position and orientation of every point in the body. It is difficult to imagine that a simpler description is possible.

These parameters are depicted in the diagram of *figure* **4.1**. Examine the diagram to convince yourself that these parameters contain a complete description of the position and orientation of an arbitrary solid The equations for the motion of the *CM* have been the subject of the previous three chapters. The time has come for us to consider the motion around the *CM*. To formulate the equations for the rotational motion around some axis, we will isolate this part of the motion by considering first those systems for which this axis is stationary. The more general motion which is the sum of the *CM* motion and the rotation will be considered later.

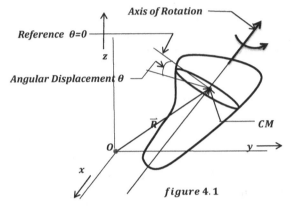

figure 4. 1

Specifying the Position and Orientation of a Solid Body

The task at hand is to determine the implications of $\vec{F} = m\,\vec{a}$ when it is applied to the angular parameters of motion. There are two distinct endeavors subsumed in this task:

1. *use ordinary geometry and algebra to formulate the equations specific to circular motion;*

2. *develop the formalism necessary to express the relations so obtained in terms of vectors.*

The first of these proceeds on the familiar grounds of analysis and serves to establish the validity of the results; the second introduces another formulation of the results in terms of vector algebra. The advantage of

the second formulation is that both the magnitude and direction of the vector quantities involved are accounted for simultaneously whereas in the first formulation, magnitude and direction must be considered separately at every step.

Recall that physics allows only those mathematical manipulations of vectors which are classified as 'vector operations' (see *sections* **3. 2. 6. 2** and **3. 2. 6. 4**), those which are independent of any coördinate system which might be introduced.

In a sense, such a formulation sidesteps any superfluous geometry and algebra and represents a strong focus on, and clarification of the physics principles involved. This can be carried further: it would be expected a priori that such a description be possible because physical quantities and processes exist without coördinate references. Consequently, one might believe in the existence of such a formalism and seek to find it. ...And even further: it can be made a logical requirement that any valid law of physics be expressible in a form freed from any coördinate system. Note that this last requirement when extended to non-Euclidean spaces was of central importance to the development of the theory of general relativity. 20-20 hindsight prompts us to snap our fingers with an air of sudden enlightenment and say, "Why didn't we think of that before?"

4.1.3 The Quantities Pertaining to This Formulation

During the course of this development, it will be expedient to use the angular parameters of motion, the angular displacement θ (defined above), the angular velocity ω and the angular acceleration α. These parameters will be viewed in the context of a point mass moving in a circle.

When a solid body is considered, then (given the axis of rotation) the values of these parameters are the same for each point of the body. Later, these parameters and the analysis of circular motion will be formulated in terms of vector algebra.

A further development of the circular motion of a point mass will be used to illustrate the basis for gyroscopic motion. If you have ever had a toy gyroscope, you already know that its motion defies your sense of 'what it should do'. When subject to an outside force, the whole system never moves in the direction in which it is pushed but rather at a right angle to it. For example, when it is free-running, the gravity force which is pulling it downward results in a motion parallel to the earth's surface and it precesses around the vertical axis which passes through the point of support.

The analysis of the circular motion of a point mass affords some insight into this apparent contradiction to both $\vec{F} = m\vec{a}$ and to our naturally acquired sense of motion. It will be seen that its motion is not a contradiction to $\vec{F} = m\vec{a}$ but rather an example of it; in regard to our naturally acquired sense of motion, we are left to resolve the apparent contradictions for ourselves.[15] Later when its motion is described in terms of vector algebra, it is seen further that the persistence of right angles is the result of a vector cross-product.

---

[15] Even when the relation between its motion and the dynamics equation is formally understood, its reaction to an externally applied force still appears unnatural. This is due at least in part to the fact that owing to the symmetry of the disk, its spinning motion is not visually obvious and thus the gyroscope looks like a solid body without moving parts. Its reactive motions however, do not parallel those of such a body.

Pervading these discussions and their generalizations to extended bodies are the definitions of the dynamic quantities which often arise in the context of circular motion. The basic quantities which will be encountered and defined along the way are:

1. *TORQUE ... the moment of force which imparts an angular acceleration to a rotating body. It is analogous to the force which imparts a linear acceleration to the linear motion of a body;*

2. *ANGULAR IMPULSE ... the time-integral of the torque. It is the analogue of the linear impulse which is the time-integral of the force;*

3. *MOMENT OF INERTIA ... the body's resistance to a change in angular motion. It is analogous to the mass which is the body's resistance to a change in linear motion;*

4. *ANGULAR MOMENTUM ... the product of the moment of inertia and the angular velocity. It is the analogue of the momentum which is the product of the mass and the velocity;*

5. *ENERGY OF CIRCULAR MOTION ... one-half the product of the moment of inertia and the square of the angular velocity. It is the analogue of the kinetic energy which is one-half the product of the mass and the square of the velocity;*

6. *WORK DONE BY A TORQUE ... the integral of the torque with respect to the angle of*

> *rotation. It is analogous to the work done*
> *by the force which is the integral of the*
> *force with respect to the distance*
> *traversed.*

The formulation of the laws for circular motion results in relations which are analogous to the laws for linear motion.

The consideration of extended bodies instead of point masses will involve the definition and computation of the moment of inertia of a solid body and a proof of the *parallel axis theorem* which relates the moment of inertia around an arbitrary axis to the moment of inertia around the axis parallel to it which passes through the *CM*.

After some examples and discussions, the angular parameters will be given vector definitions and the equations for general acceleration will be obtained in vector form. The overall intent of this chapter is not only the formulation of these equations but also the development of an understanding of each term in them.

4.2   CIRCULAR MOTION OF A POINT MASS

4.2.1 The Scaler Form of the Angular Parameters

Consider a point mass *m* moving in a circular path of radius *r* as shown in *figure* **4. 2**. Let the *z*-axis be the axis of rotation so that the motion takes place in the *x,y*-plane. The task at hand is subsumed in the larger task of describing the orientation of an extended body in terms of its angular displacement. We therefore want to relate the angle of rotation around the *z*-axis to the position of the mass on the circumference of the circle. (Recall our earlier decision that the equations of motion be formulated in terms of this angle.) The following remarks pertain to the scaler definitions of angular speed and angular acceleration.

The basic relations are simple. The distance s along the circumference from the reference direction (taken as the x-axis) is

$$s \; = \; r\theta .$$

where the angle θ is the angular displacement.

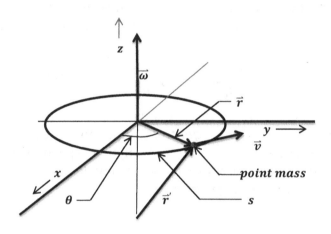

figure 4.2

Circular Motion of a Point Mass

The speed of the mass and the rate of change of the displacement angle can be related by taking the time-derivative of the above relation:

$$\frac{ds}{dt} \; = \; r\,\frac{d\theta}{dt},$$

where all the quantities are scalers and r is constant since the motion is circular. The derivative on the left is the rate at which distance is covered along the circumference, i.e. the speed of the mass. The derivative on the right is the rate of change of the angle θ, measured in the number of radians per second. This quantity is ever-present when considering circular motion. It is called the **angular velocity** and is usually designated ω. If $\vec{v}$ is the velocity, then the last equation can be written

$$|\vec{v}| \quad = \quad r\omega.$$

Another relation of interest relates the linear and angular accelerations. It is obtained by taking the time-derivative of the above equation

$$\frac{d|\vec{v}|}{dt} \quad = \quad r\frac{d\omega}{dt} \quad = \quad r\frac{d^2\theta}{dt^2}.$$

The derivative on the left is the rate of change of the speed and hence is the **tangential** component of the acceleration. The time-derivative of the angular speed is called the **angular acceleration** and is usually designated α. The last equation can therefore be written:

$$a_t \quad = \quad r\alpha.$$

Thus in the scaler model, it is seen that the angular speed can replace the speed and the angular acceleration can replace the t

angential component of the acceleration. This approach has failed to account for the radial component of acceleration: it must be taken care of separately. (Details such as this are automatically taken care of in the vector formulation.)

To this end, consider the vector diagram in *figure* 4.3. The mass is shown at a time t and at a time Δt later:

figure 4.3.*b* shows the velocity vectors at these two times with their tails together so they can be subtracted graphically. From the diagram, it is clear that

$$|\Delta\vec{v}| = |\vec{v}|\Delta\theta.$$

Note that $\Delta\vec{v}_t = 0$ because the speed is constant. Therefore the change in the velocity vector is in the (negative) radial direction. (To see this from the diagram, reference the vector $\Delta\vec{v}$ at the position of the mass.) To find the radial component of acceleration, divide by Δt and take the limit as $\Delta t \to 0$:

$$a_r = -|\vec{v}|\frac{d\theta}{dt} = -|\vec{v}|\omega$$

Since the speed $|\vec{v}|$ and the angular speed ω are proportional, a_r is usually written in terms of one of them, i.e. in one of the two forms

$$a_r = -\omega^2 r = -\frac{|\vec{v}|^2}{r}.$$

The radial component of the acceleration changes the direction of motion in accordance with the last relation. It is called the centripetal (center seeking) acceleration and is non-zero whenever the path of motion is anything other than a straight line.

Finally, the force necessary to produce the circular motion of the mass is as follows:

1. **the tangential component of force is**
$$F_t = ma_t = mr\alpha.$$
2. **the radial component of force is**
$$F_r = ma_r = -m\omega^2 r = -\frac{m|\vec{v}|^2}{r}$$

The **tangential component** F_t changes the speed of the mass in keeping with the work-energy law. The **radial** *component F_r* changes the direction of motion by keeping the mass on the circumference of the circle. It is called the *centripetal (center-seeking) force.*

4.2.2   The Vector Form of the Angular Parameters

We seek another formulation of the same kinematics as described in the previous section. The issue is to find a vector representation of the angular parameters ω and α which relate through vector operations to the velocities and accelerations found above. The kinematic relations expressed in vector form will be freed from the reference to any particular coördinate system.

Circular motion in the xy-plane does not allow for the singling out of a particular direction in the plane for a vector parameter which is to represent the motion: any attempt to do so would be unreasonable since the nature of circular motion renders all directions in the plane of motion equivalent. Also one of the most important applications of the angular parameters is the description of the rotational motion of solid bodies: it is in this context that the angular versions of displacement, speed and acceleration refer to the entire body as a unit. Recall that one of the initial motivations was to avoid a detailed description of the body, in particular that of specifying the parameters for each point in its interior.

In keeping with the above remarks, it is reasonable to attempt to find a representation in which a *constant* angular velocity is represented by a *constant* vector. The previous remarks indicate further that such a representation cannot have a component in the xy-plane because any such component would single out a preferred direction. It follows that if such a representation is possible, the vector representing the angular velocity must point in the **z**-direction.

There are two rotational possibilities: clockwise and counter-clockwise (looking from above). There are also two possibilities for the direction of the angular velocity vector: the positive **z**-direction and the negative **z**-direction. Since there is no apparent reason to prefer one of the two possible definitions over the other, the following assignment is (arbitrarily) chosen: when the thumb of your right hand points in the direction of the angular velocity vector $\vec{\omega}$, your fingers when slightly curved, point in the direction of rotation.

Exercise: Refer to *figure* 4.2 and check the above definition for the direction of rotation.

At this point, the above definition is on trial: it remains to show that it leads to a consistent vector representation for all the results obtained in the preceding scaler analysis. First note that if we use the above definition of the angular velocity vector, then it follows that the relation between $\vec{v}$ and $\vec{\omega}$ is

$$\vec{v} \;=\; \vec{\omega} \times \vec{r}$$

where $\vec{r}$ is the position vector of the mass.

Note two things:

1. *the velocity vector produced by the indicated cross-product is correct in both magnitude and direction (refer to figure 4.2);*

2. *we got a freebee! If $\vec{r}'$ is a vector from any point on the axis of rotation to the mass, it will work as well in the above relation since the cross-product singles out the component normal to the axis.*

The first kinematic relation of the previous section has therefore been successfully written as a vector relation.

Since the cross-product represents the **vector velocity** (not just the speed), the **acceleration vector** should be obtainable by taking the time-derivative:

$$\vec{a} \;=\; \frac{d\vec{v}}{dt} \;=\; \frac{d(\vec{\omega} \times \vec{r})}{dt}$$

$$\vec{a} \;=\; \vec{\omega} \times \frac{d\vec{r}}{dt} \;+\; \frac{d\vec{\omega}}{dt} \times \vec{r}$$

$$\vec{a} \;=\; \vec{\omega} \times \vec{v} \;+\; \vec{\alpha} \times \vec{r}$$

If the previous expression for the velocity $\vec{v}$ is used, then

$$\vec{a} \;=\; \vec{\omega} \times (\vec{\omega} \times \vec{r}) + \vec{\alpha} \times \vec{r}$$

In this last expression the first term on the right is the **centripetal acceleration** (note that it points toward the axis of rotation); the second is the **tangential acceleration.** The angular acceleration α points in the **z**-direction because it is the difference (derivative, $lim_{\Delta t \to 0} \frac{\Delta \omega}{\Delta t}$) of two angular velocity vectors both of which point in the **z**-direction.

Note that in its usual form the dynamics equation $\vec{F} \;=\; m\vec{a}$ does not reference any coördinate system. The previous development has just shown that the application of this equation to the motion of a point mass **m** moving in a circle of radius **r** around a given axis can be described by the vector equation:

$$\vec{F} = m\vec{a}$$

$$\vec{F} = m[\vec{\omega} \times (\vec{\omega} \times \vec{r})] + m(\vec{a} \times \vec{r}).$$

Since the last equation does not reference any coördinate system, any coördinate system can be chosen to analyze this circular motion problem by expressing the vectors and cross products in that system. The result is the equation of motion expressed in the coördinates chosen.

(Notice that once the vector velocity was expressed as a vector operation (a cross-product) involving the angular velocity vector and a vector from the axis of rotation to the mass, the acceleration could be found by straightforward differentiation: there was no need to separate out the individual components and apply a new analysis to each of them. This is much more satisfying than the scaler approach.)

Exercise: Show that the first term in the above expression for the force is the centripetal force and the second is the tangential force as determined in the scaler analysis.

4.2.3 The Basis of Gyroscopic Motion

A gyroscope is a device containing a symmetric disk that spins on an axis. It is not only a curious toy but also a sensing device used for navigation as well as part of a feedback system used to stabilize ships, planes and spacecraft. Its curious nature derives from the fact that it always responds to an external stimulus by moving at right angles to the applied force. This offends our sense of the natural motion of objects whose acceleration is usually in the same direction as the applied force ala $\vec{F} = m\vec{a}$. Hence, it also produces a paradox in that it seems to contradict the law of dynamics.

The analysis of gyroscopes is much too long and intricate for presentation here, but the following tidbit affords insight into its impish move-at-ninety-degrees-from-the-way-it's-'supposed'-to behavior. It will indicate that its motion is not a contradiction of, but rather a compliance with $\vec{F} = m\vec{a}$.

We will examine what happens when a point mass moving in a circle is acted upon by an outside force. (Actually, matters are simplified by giving it a sharp blow, i.e. an impulse. There is no loss of generality in doing this.)

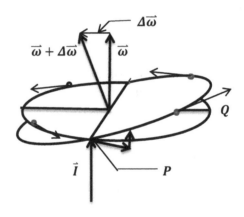

figure 4.4

Reaction of a Gyroscope to an Applied Impulse

In *figure 4.4* the mass is depicted at five different positions during its motion. The first three show it as it moves in a circle with constant speed v in the xy-plane. These are the three positions on the left side of the diagram. When it reaches the third position at **P** however, it is subject to an impulse **I** acting in the positive **z**-direction.

The effect of the impulse is to impart to the mass, a velocity component in the **z**-direction as shown in the figure. Consequently, the mass continues its circular motion in a plane tilted around the diameter that ends at **P**. The angle φ of the tilt is found from (see ***figure* 4.4**)

$$tan(\varphi) \quad = \quad \frac{I_z}{mv}.$$

mass with that of a circular disk spinning in the **xy**-plane. It would seem that imparting the impulse to the disk at **P** would tilt the plane of motion around the diameter that ends at **Q**. But this is not so; it tilts around the diameter that ends at **P** for the same reason as that presented for the point mass. Thus it moves at right angles to its 'expected motion'.

Later it will be seen that the vector form of the dynamics equation produces the same result when the rotating part of the gyroscope is a solid disk or a ring.

The analysis is as follows. The impulse-momentum law applies:

Impulse = Change in Momentum

$$I_z \quad = \quad mv_z \; - \; mv_{z_0}$$

$$I_z \quad = \quad mv_z \; - \; 0$$

$$\Rightarrow \quad v_z \quad = \quad \frac{I_z}{m}.$$

4.2.4   The Torque Equation

The subject of moments was introduced in **section** 2.2 in connection with the balancing of a see-saw. Recall that moments refer primarily to rotational motion: besides the particular physical quantity involved, the distance from the center of rotation to the point of interest, called the moment arm also enters into consideration. It was determined that the see-saw was in a state of balance when

$$MR \quad = \quad mr \,,$$

where M and m are masses whose respective distances from the fulcrum are R and r. The following illustrates that the conclusion follows from $F = ma$ and thereby establishes the role of moments as the factors that empower circular motion.

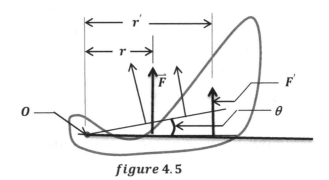

figure 4.5

Circular Motion and the Moment of Force (Torque)

Consider an arbitrary solid pinned at point O. A force of magnitude F is applied at a point which is a distance r from the pivot. Suppose that the only resistance to motion is the body's own inertia. Let the force be applied in such a way that it is always normal to the moment arm (see *figure* 4.5). When the body has rotated through some arbitrary angle θ, the kinetic energy can be found from:

Work = Change in Kinetic Energy

$$\int_0^\theta \vec{F} * d\vec{s} \;=\; T_f - T_i$$

$$Fr\theta \;=\; T_f - 0$$

$$\Rightarrow \quad T_f \;=\; Fr\theta$$

Now perform this operation again using a second force F' applied at a distance r' from the pivot. Let the magnitude of F' be adjusted so that the motion of the body is identical to its previous response to the force F. The final kinetic energy (when the displacement angle is θ) is

$$T'_f \;=\; F'r'\theta .$$

Since the two motions are identical,

$$T_f \;=\; T'_f$$

$$\Rightarrow \quad Fr \;=\; F'r'.$$

The moment of force (Fr or $F'r'$ *above*) is called the **torque.** This analysis shows that identical responses result from equal torques. It can therefore be concluded from the above that

> *the quantity that empowers circular motion around a given axis is the moment of force or torque around that axis.*

This formally justifies the 'reasonable' conclusion drawn from the see-saw analysis.[16]

---

[16] See last paragraph of *section* 2.2.1.

The torque equation proceeds from $\vec{F} = m\vec{a}$ by considering only the tangential component of F:

$$F_t \;\; = \;\; ma_t$$

$$\tau \;\; = \;\; F_t r \;\; = \;\; mra_t$$

$$\tau \;\; = \;\; (mr^2)\alpha.$$

The final expression above is the torque equation. It is the circular motion analogue of $F = ma$.

The mass m in the dynamics equation, which is the coefficient of the linear acceleration and whose complete name is the **inertial mass** is said to be the resistance to a change in motion because, for example, if the mass is doubled, then the force must be doubled to result in the same motion. Analogously, the quantity mr^2 in the last expression, which is the coefficient of the angular acceleration in the torque equation is called the **moment of inertial mass** or more briefly, the **moment of inertia** and is usually designated I. It is the resistance of the body to a change in rotational motion in the same sense that mass is the resistance to a change in linear motion.

We therefore have that the moment of inertia of a point mass a distance r away from the axis of rotation is

$$I \;\; = \;\; mr^2.$$

We also have the form of the torque equation which is analogous to the dynamics equation

$$F = ma,$$

$$\tau \;\; = \;\; I\alpha$$

Note that this applies only to a point mass: the moment of inertia for an arbitrary body depends on the shape, size, and mass distribution of the body and on the axis of rotation. More about this will be said in the following section.

4.3   THE MOMENT OF INERTIA OF SOLID BODIES

4.3.1   The Relation between Inertial Mass and the Moment of Inertia

The above relation for the moment of inertia of a point mass is sufficient to find the moment of inertia of any solid body if one conjectures reasonably (and correctly) that the moments of inertia of the pieces of a body add together to produce the moment of inertia of the whole body. Finding it is essentially a math problem: you break the body up into differential size pieces, calculate the contribution of a typical piece to the moment of inertia, then add them all together (integrate).

Although you are not expected to perform this operation, several examples of the calculations involved will be presented in the following section. Before continuing however, it is the point of the following discussion to prove that the moment of inertia of a solid body is the algebraic sum of the moments of inertia of its differential pieces.

Figure 4.6 depicts a bent rod with masses m and m' attached. One end of the rod is pinned at the origin of the xy-plane and the assembly has angular acceleration α around the origin. Let r and r' be the respective distances of the masses from the origin, and a and a', their respective linear accelerations

Applying the dynamics equation to m, the force and torque which must be delivered to m are

$$F \quad = \quad ma$$

$$F = mr\alpha$$

$$\tau = mr^2\alpha$$

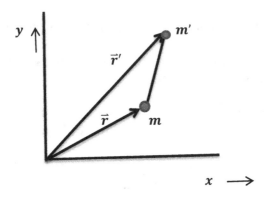

figure 4.6

Moment of Inertia of a Solid Body

And similarly for **m'**:

$$\tau' = m'r'^2\alpha$$

The net torque τ_o that must be delivered is the sum of the two above torques:

$$\tau_o = \tau + \tau'$$

If the last equation is compared to the torque equation (1), it is seen that the quantity in parentheses is the moment of inertia of the assembly. But this same quantity is also the simple algebraic sum of the moments of inertia of the individual masses.

> *the total moment of inertia of a solid body is the algebraic sum of the moments of inertia of its parts.*

4.3.2   The Computation of the Moment of Inertia

The computation of the moment of inertia is primarily a problem in integration which you are not expected to carry out. These few examples are included as an illustration of the process and to determine some of the moments of inertia which will be needed in the sample problems of *Chapter V.*

In addition, the parallel axis theorem will be proven in the next section. It states that the moment of inertia of a solid body around an arbitrary axis is the moment of inertia of the *CM* around that axis plus the moment of inertia of the body around the parallel axis which passes through the *CM*. It is therefore necessary to consider only moments of inertia around axes through the *CM* since all others can be found from these by use of the theorem.

We will stick to two dimensions. While the extension to three dimensions involves interesting advances in mathematics, namely the use of vector-matrix algebra, an excursion into this topic represents too large an undertaking at this point and would tend to mask the physics involved.

The approach followed here is to avoid these questions, leaving them for a more mathematically oriented presentation of this material. However, several remarks which allude to this extension are incorporated to indicate when our restricted version touches on issues which are normally addressed in a description which makes use of this mathematics.

4.3.2.1 The Moment of Inertia of a Disk

Consider a disk of mass **M** and radius **R**. Then the mass per unit area is

$$\sigma \;=\; \frac{M}{\pi R^2}$$

The disk is depicted in **figure 4.7**. In the middle of the wedge-shaped piece is a differential square-like 'block' whose dimensions are $rd\theta$ in the angular direction and dr in the radial direction. To find the moment of inertia **I** of the disk, the contribution from each of these differential pieces must be added.

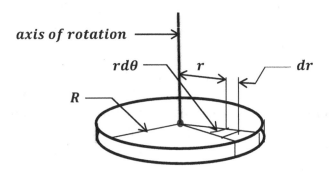

figure 4.7

Moment of Inertia of a Disk

The area **dA** of this piece is

$$dA \;=\; rd\theta dr \;;$$

its mass **dm** is the mass per unit area times the area,

$$dm \;=\; \sigma rd\theta dr;$$

and its contribution to the moment of inertia **I** is

$$dI \;=\; r^2 dm$$

$$dI \;=\; \sigma r^3 d\theta dr .$$

If these contributions are added together from $r = 0$ to $r = R$, the result is the differential contribution of the (pie-shaped) wedge: then if the contributions from these wedges are added together from $\theta = 0$ to $\theta = 2\pi$, the result is the moment of inertia of the whole disk.

$$I \;=\; \int dI$$

$$I \;=\; \sigma \int_0^{2\pi} \left[\int_0^R r^3 dr \right] d\theta$$

$$I \;=\; \sigma \int_0^{2\pi} \frac{R^4}{4} d\theta$$

$$I \;=\; \frac{2\pi \sigma R^4}{4}$$

Notice that $M = \sigma \pi R^2$. Then the moment of inertia of the disk is

$$I \;=\; \frac{1}{2} MR^2 ,$$

4.3.2.2 The Moment of Inertia of a Thin Rod

Let L be the length of the rod and M, its mass. Then the mass per unit length μ is

$$\mu \;=\; \frac{M}{L} .$$

The rod is shown in *figure* 4.8. The differential mass of length dx is the mass per unit length times the length,

$$dm = \mu dx:$$

its contribution to the moment of inertia is

$$dI = \mu x^2 dx.$$

These contributions must be added together from $x = -\frac{1}{2}L$ to $x = +\frac{1}{2}L$:

$$I = \int_{-\frac{L}{2}}^{+\frac{L}{2}} \mu x^2 \, dx$$

$$I = \mu \left[\frac{L^3}{24} - \left(\frac{-L^3}{24} \right) \right]$$

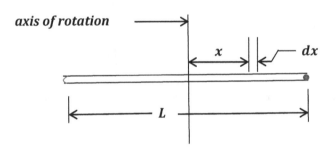

figure 4.8

Moment of Inertia of a Thin Rod

Note that $M = \mu L$. Then

$$I = \frac{ML^2}{12}$$

The calculations for all moments of inertia are essentially the same: add the moments of inertia for each differential piece.

4.3.2.3 The Parallel Axis Theorem (Lagrange, 1783)

Whenever the circular motion of a solid body is to be determined, it is necessary to know the moment of inertia around the axis of rotation. If the axis does not go through the **CM**, the moment of inertia can be related to the moment of inertia around the parallel axis that does. The following theorem known as the parallel axis theorem provides this relation.

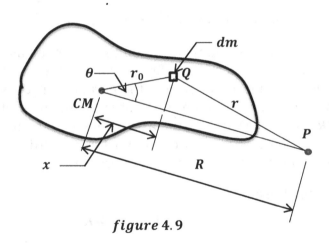

figure 4.9

The Parallel Axis Theorem

The proof is straightforward. Refer to *figure* 4.9 which is a diagram depicting such a situation. An arbitrarily shaped plate is shown with its **CM**. Let the axis through the **CM** be

· normal to the plane of the body and parallel to the axis through **P** around which the body is rotating. Let I_o be the moment of inertia around the axis through the **CM** and I, the moment of inertia around the axis through **P**. Let **R** be the distance from **O** to **P** and align the **x**-axis so its origin is at the **CM** and **x** is measured along the line **OP**.

Consider a typical differential mass **dm** at some point **Q** and let r_0 be its distance from the **CM** and **r**, its distance from **P**. From the law of cosines (refer to **figure** 4.9).

$$r^2 \quad = \quad r_0^2 \; + \; R^2 \; - \; 2r_0 R\cos(\theta)$$

Multiply this equation by **dm,** integrate and collect the results:

$$\int r^2 dm \; = \; \int r_0^2 \, dm \; + \; R^2 \int dm \; - \; 2R \int x dm \, ,$$

where **x** has been substituted for $r_0 cos(\theta)$.

Consider each term: the term on the left is the moment of inertia I around **P**; the first term on the right is the moment of inertia I_o around the parallel axis through the **CM**; the second term on the right is the moment of inertia MR^2 of the **CM** around the axis through **P**; the integral in the last term is the expression for the **x**-coördinate of the **CM** in the **CM** system and is therefore zero.

The final result is

$$I \quad = \quad I_0 \; + \; MR^2$$

In words,

> *the moment of inertia around an arbitrary axis of rotation is equal to the moment of inertia of the CM around that axis plus the moment of inertia around the parallel axis through the CM.*

As an example, consider the moment of inertia of a disk of radius R and mass M when it is rotated around a point on its rim (rather than around an axis through its center). The moment of inertia from the parallel axis theorem is

$$I = I_0 + MR^2$$

$$I = \frac{1}{2}MR^2 + MR^2$$

$$I = \frac{3}{2}MR^2.$$

4.4 THE MOTION OF RIGID BODIES

4.4.1 The Torque Equation for Extended Rigid Bodies

The result of *section* 4.3.1 was that in a system composed of two rigidly connected point masses, the moment of inertia of the assembly was the algebraic sum of the moments of inertia of the individual masses. If the two masses are considered as two differential pieces of an extended body, then repeated application of this result (i.e. integrating over the whole body) implies that the net torque on the body satisfies the equation

$$\tau = I\alpha.$$

The only task remaining to make this equation useful is to show that the torque on the left is the **net external** torque on the system. It must be shown that for any configuration of internal forces which satisfy the action-reaction law, the net torque from the internal forces around any arbitrary point (axis) is zero.

The masses depicted in *figure* 4.10 can be thought of as two differential pieces of a rigid body each exerting a force on the other in compliance with the action-reaction law and thus producing torques around the arbitrary point **P.**

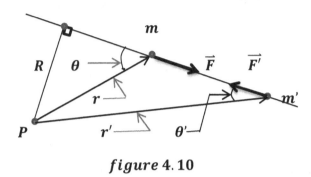

figure 4. 10

The Torque Equation for Extended Bodies

The result needed is obtained by computing the net torque around **P**:

$$\tau_P \;=\; |\vec{F}||\vec{r}|\,sin(\theta) \;-\; |\vec{F'}||\vec{r'}|\,sin(\theta')\,.$$

Note that

$$|\vec{r}|\,sin(\theta) \;=\; |\vec{r'}|\,sin(\theta') \;=\; R$$

so that the sum of the torques around **P** is

$$\tau_P \;=\; |\vec{F}|R \;-\; |\vec{F'}|R$$

$$\tau_P \;=\; (|\vec{F}| - |\vec{F'}|)R$$

As long as the forces **F** and **F'** are equal and opposite, and act along the line joining **m** and **m'**, the torque from these internal forces around any arbitrary point **P**, add to zero. Therefore, the left side of the torque equation refers to the **net external torque.** This is the result we were after.

(Note however, that in order to obtain this result, it was necessary not only that $\vec{F}$ and $\vec{F}'$ be equal and opposite but also that they acted along their line of centers. The electro-dynamic forces of two charges moving in each other's fields do not satisfy the second of these conditions. But we are considering a **rigid body**: in this case, there can be no relative motion of the mass points in question and hence, even if the masses are charged, there are no off-the-line-of-centers forces in play.)

Exercise: Even though it is not an issue here, prove that when $\vec{F}$ and $\vec{F}'$ are equal and opposite but do not act along the line of centers, that the torque equation cannot be applied as above. Illustrate that there exist points around which the torque due to internal forces is non-zero. (Hint: draw a diagram analogous to that of *figure* **4.10** which depicts the situation described here.)

4.4.2 The Vector Form of the General Torque Equation for Point Masses

Before the derivation of the torque equation, a few remarks may afford an overview of what we are doing. Recall that we have defined moments, the angular velocity vector, the angular acceleration vector and now the torque vector in the most reasonable fashion available to us as we went along. We were in the process of **defining** a set of parameters which could be used to express the equations of circular motion.

There are many possible ways to define sets of parameters and any such set could be used to the same purpose. The tacit belief however as we went through these definitions was that the simplest definitions would lead to the simplest form of the equations.

Also recall that many times in algebra, you have said, 'let x equal so-and-so; then the above equation written in terms of x is...'. We are doing the same thing here: 'let $\vec{\omega}$ point in the positive z-direction, then the velocity can be written as $\vec{\omega} \times \vec{r}$; it has the correct magnitude and the correct direction...'. This is part of assembling a new set of variables in terms of which our results about the motion of bodies. can be expressed. We are simply rewriting $\vec{F} = m\vec{a}$ in terms of the parameters we deem pertinent to the description of circular motion. It should be noted further that there is some 20-20 hindsight involved here: these parameters have been used for a long time, have been scrutinized by many workers and have proven themselves to be valuable descriptors.

The only difference between the following derivation and the algebraic analogy cited above is that we are singling out a particular component of the force by taking cross-products. The results then, while still valid are a restricted version of the original equation. This is purposefully so because we want to focus attention on those factors of the original equation which pertain to motion around an axis or point.

The point is that the following derivation is not some magic trick. It is not even some complex relation that some genius 'noticed' would lead to an economical analysis of circular motion: it is simply the culmination of our efforts to gather a good set of parameters. Enough isolated facts have been gathered at this point to see if they can in fact be made into a self-consistent and useful application of $\vec{F} = m\vec{a}$ to circular motion.

Cross-products are tailor made for this application to circular motion: the sine of the angle between the factors in the product singles out the normal component of one of the two vectors which is particularly suited to the consideration of moments. Since the torque is defined as the moment of force, the torque **vector** can be written as

$$\vec{\tau} \;=\; \vec{r} \times \vec{F}$$

as long as the direction is compatible with our other vector definitions.

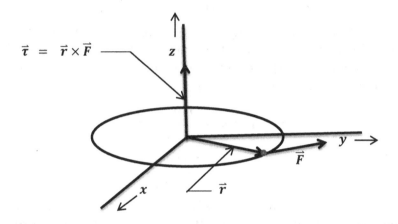

figure 4.11

The Vector Formulation of the Torque Equation for Point masses

The diagram in ***figure* 4.11** depicts the general situation currently under consideration.

Once again, we start with $\vec{F} = m\vec{a}$ for a point mass: the torque equation is found by crossing the position vector $\vec{r}$ into it:

$$\vec{F} \;=\; m\frac{d\vec{v}}{dt} \;=\; \frac{d\vec{p}}{dt}$$

$$\vec{r} \times \vec{F} \;=\; m \left(\vec{r} \times \frac{d\vec{v}}{dt} \right)$$

To get this relation in a form that looks more analogous to the linear equation (the first one above), we rewrite the cross product a little differently. Consider the time-derivative

$$\frac{d(\vec{r} \times \vec{v})}{dt} \;=\; \frac{d\vec{r}}{dt} \times \vec{v} + \vec{r} \times \frac{d\vec{v}}{dt}.$$

The first term on the right is the velocity vector crossed with itself which is zero (the cross product of any vector with itself is zero). The second term is the same as the parentheses in the previous equation. Therefore,

$$\vec{r} \times \vec{F} \;=\; \frac{d(\vec{r} \times m\vec{v})}{dt}$$

$$\vec{r} \times \vec{F} \;=\; \frac{d(\vec{r} \times \vec{p})}{dt}$$

The quantity in parentheses in the last equation is the **moment of momentum.** It is called the **angular momentum** and is usually designated $\vec{L}$. The general torque equation therefore is

$$\vec{\tau} \;=\; \frac{d\vec{L}}{dt}.$$

It is analogous to the linear equation

$$\vec{F} \;=\; \frac{d\vec{p}}{dt}.$$

Note that no assumptions have been made up to this point: all that has been done is to cross the position vector into the dynamics equation and rewrite the result in terms of the torque and a quantity we defined as the angular momentum. The above torque equation is the general result for a point mass.

Before taking advantage of the similarity in form between these two equations, the result will be established for arbitrary extended bodies. In the process, the results become automatically restricted to the motion of rigid bodies. The reason for this is that for an extended fluid body, the angular parameters and the moment of inertia have no meaning in general: the differential pieces of a liquid or gas generally do not have the same angular velocity or angular acceleration around an axis or point. Hence the use of the angular parameters in these cases is pointless: fluids are therefore analyzed using the above equation for each differential volume. On the other hand, the angular parameters apply to every point of a rigid body affecting a considerable simplification of the analysis of their motion.

In the most general case of the motion of a rigid body, the moment of inertia no longer enjoys the status of a simple scaler quantity: the physics of the motion and the requirement that the vector directions be compatible conspire to make it into a 3×3 matrix. This is not the case if the motion is restricted to two dimensions. This restriction is imposed so that we may avoid issues of vector-matrix algebra and focus on the physics principles involved. For the remainder of this chapter, considerations are confined to the 2-dimensional motion of rigid bodies.

4.4.3   The Torque Equation for the 2-Dimensional Motion of Rigid Bodies

For the sake of specificity, let the motion take place in the xy-plane. Then all vectors pertaining to forces, accelerations, velocities, positions and momenta are confined to that plane while all vectors pertaining to angular parameters (including torque and angular momentum) will be in the z-direction.

First consider the angular momentum:

$$\vec{L} = \vec{r} \times \vec{p}$$

$$\vec{L} = m(\vec{r} \times \vec{v})$$

$$\vec{L} = m[\vec{r} \times (\vec{\omega} \times \vec{r})]$$

We wish to show that $\vec{L} = I\vec{\omega}$.

Since $\vec{L}$ and $\vec{\omega}$ both point in the z-direction, (check the directions of the above cross-products), it can be concluded that the directions implied are correct. Now the magnitude: $\vec{\omega}$ and $\vec{r}$ are normal to each other, hence, the magnitude of the cross-product in parentheses is

$$|\vec{\omega} \times \vec{r}| = |\vec{\omega}||\vec{r}|$$

Furthermore, this cross-product is normal to the position vector. Hence,

$$|r \times (\omega \times r)| = |\vec{r}||\vec{\omega} \times \vec{r}|.$$

Combining the last two equations,

$$|r \times (\omega \times r)| = |\vec{r}|^2|\vec{\omega}|.$$

Multiply by m and use the moment of inertia for a point mass $(I = m|\vec{r}|^2)$, and the equation becomes

$$\vec{L} = I\vec{\omega}$$

which is correct in both magnitude and direction.

Now consider the point mass to be a differential piece of a solid body and add the contributions of all these pieces. It has been shown already (*section* **4.4.1**) that the moments of inertia add. Also, since all the torques point in the same direction, the magnitude of their vector sum is just the sum of their magnitudes. It follows that the relation remains in tact under this addition (integration), and

> *the torque equation for the 2-dimensional motion of a rigid body is*

$$\vec{\tau} \;=\; \frac{d\vec{L}}{dt}$$

$$\vec{\tau} \;=\; \frac{d(I\vec{\omega})}{dt}.$$

4.5   THE DYNAMIC IMPLICATIONS OF THE TORQUE EQUATION

4.5.1   Angular Impulse and Angular Momentum

4.5.1.1 The Derivation of the Relation

We proceed now as in the case of linear motion to deduce the simplest relations that follow directly from the torque equation. The analogue of *section 1.2.3* is the integration of the equation as it stands:

$$\vec{\tau} \;=\; \frac{d\vec{L}}{dt}$$

$$\int \vec{\tau} dt \;=\; \Delta\vec{L}$$

$$\int \vec{\tau}\, dt \;=\; \vec{L}_f \;-\; \vec{L}_i$$

The integral on the left is called the **angular impulse** and is usually designated $\vec{J}$. This relation in words is

Angular Impulse = Change in Angular Momentum

Note two things:

1. *no assumptions about the motion were made in the above. The result applies to any motion even if it is in a straight line;*

2. *the angular momentum is a vector that points in the direction of the axis of rotation.*

An immediate sequel to the angular impulse-momentum law is the...

> *Conservation of Angular Momentum: if the net torque around any point P is zero then the angular momentum around P is conserved, i.e. it is a constant vector throughout the motion.*

4.5.1.2   The Vector Description of the Basis for Gyroscopic Motion

In **section 4.2.3** the basis for the strange reactions of a gyroscope to the application of an external force was explained using the linear equations of motion. We are now in a position to demonstrate how the vector form of the torque equation applies to the same issue.

Refer to *figure 4.12* and recall that the point mass m was initially moving in a circle with constant speed v and that an impulse I was delivered to the mass at point P. The normally expected response is for the plane of motion to rotate around the diameter at Q, but instead, it rotated around the diameter at P. It is shown below that the vector formalism we are pushing here arrives at the same result. The understanding of the physics comes from the previous analysis; the vector formalism offers no new insights in this regard. Its value comes from the fact that it can be easily and mindlessly applied.

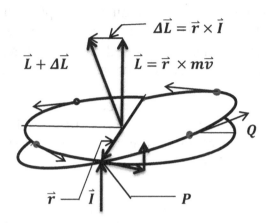

figure 4.12

The Vector Description of the Basis for Gyroscopic Motion

To obtain the result, use is made of the angular impulse-momentum relation:

Angular Impulse = *Change in Angular Momentum*

$$\vec{J} = \Delta \vec{L}$$

$$\vec{r} \times \vec{I} = \Delta \vec{L}$$

The right-hand rule for the cross-product indicates that the change in angular momentum is parallel to the diameter at **Q** as shown. Reference to the diagram of *figure* 4.12 shows that the final angular momentum is displaced from the **z**-axis by an angle φ, which satisfies the equation

$$tan(\varphi) \quad = \quad \frac{I}{mv}.$$

This is the same result as that obtained in **section 4.2.3**.

Notice finally that the same analysis applies to any mass rotating in the *x,y*-plane as long as the appropriate moment of inertia is used to find $\vec{L}$. This includes the solid disk or ring of a real gyroscope. The vector description exhibits the essential factors which govern gyroscopic motion: the interplay between angular impulse and angular momentum. It is the cross-product in the above equation that is at the bottom of the move-at-right-angles-to-the-expected-reaction behavior.

4.5.1.3   Examples of the Conservation of Angular Momentum

Example 1

Consider a point mass *m* traveling in the *x*-direction along the line *y* = *a* and not acted upon by any external force as shown n *figure* **4.13.a**. In the absence of any external force or torque, both linear and angular momenta are conserved. To demonstrate that angular momentum is conserved around the origin even in this case where the motion is linear, note that $\vec{r}$ and $\vec{p}$ are both in the *xy*-plane so that their cross product is in the (negative) *z*-direction. It follows that the direction of $\vec{L}$ cannot change. Concerning its magnitude, consider

$$|\vec{L}| \quad = \quad |\vec{r}||\vec{p}|\, sin(\theta).$$

The magnitude of $\vec{p}$ is constant because linear momentum is conserved; although $sin(\theta)$ and the magnitude of $\vec{r}$ both vary, their product is a constant and is equal to the distance from the origin to $y = a$. It follows that $\vec{L}$ is constant in both magnitude and direction and is therefore conserved.

Example 2

Consider a rubber band pinned at the origin of a frictionless table top with its other end attached to a mass m which is given some initial velocity $\vec{v}_o$. Sometimes the rubber band will stretch, at other times it will sag. There are unknown forces at the pin. In general, the mass will exhibit very complex gyrations under these circumstances. (Refer to *figure* **4. 13. b.**)

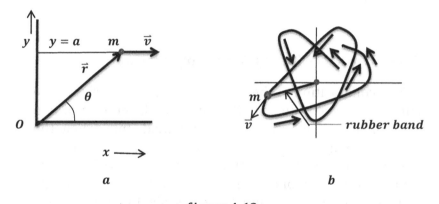

figure **4. 13**

The Conservation of Angular Momentum

Linear momentum is not conserved because of the unknown forces at the pin. The angular momentum around any point **except the origin** is not conserved either because the forces at the pin have a non-zero torque around all other points in the plane.

Now consider the origin. The torque around the origin from the unknown forces is zero because the moment arm is zero. Also the forces exerted by the rubber band on the mass are radially directed and therefore have no torque around the origin. Hence, the angular momentum $\vec{L}$ around the origin is conserved.

This conclusion is a rather strong result: it means that no matter where the mass may be at some given instant, the position vector $\vec{r}$ crossed with the momentum vector $\vec{p}$ (or the velocity $\vec{v}$) is a constant. This establishes a relation between the position and velocity of the mass. This relation which is so easily extracted from the theory is not at all obvious.

4.5.2 Work and Kinetic Energy of Rotation

The decomposition of the kinetic energy of a body into the sum of two separate contributions, one from each of two components of velocity can be justified only when the velocity components are orthogonal. Remember, kinetic energy is a *scaler* quantity and cannot be broken down into (vector) components. The fact that the kinetic energy can be written as the sum of two (scaler) quantities, one from each component of the velocity is a happy accident which proceeds from Pythagoras' theorem.

If $\vec{v}'$ and $\vec{v}''$ are two orthogonal components of the velocity $\vec{v}$ of a moving mass **m,** Pythagoras' theorem asserts

$$|\vec{v}|^2 \quad = \quad |\vec{v}'|^2 \ + \ |\vec{v}''|^2.$$

If this relation is multiplied by $\frac{m}{2}$, then the resulting equation

$$\frac{1}{2} mv^2 = \frac{1}{2} mv'^2 + \frac{1}{2} mv''^2$$

exhibits the total kinetic energy (on the left) as the sum of two terms, one from each of the components of the velocity vector.

These statements apply when the 2-dimensional motion of a rigid body is decomposed into the orthogonal radial and angular components. We are therefore justified in singling out one of these terms, the one associated with the angular direction: it is referred to as the **kinetic energy of rotation.** It will now be shown that the work-energy relation that results from the torque equation is

$$\int \vec{\tau} \, d\theta = \frac{1}{2} I(\omega_f^2 - \omega_i^2).$$

Since we are considering only that component of $\vec{F}$ which is normal to $\vec{r}$, there is no loss of generality if we consider pure circular motion. **Figure 4.14** shows an extended body pinned at the origin. It is free to rotate in the *xy*-plane; a force $\vec{F}$ is applied at a distance r from the origin in such a way that it always points in the direction of motion.

Consider first that there is a point mass m instead of the extended mass shown. It will be seen that the result for the extended body follows easily from the point mass result.

The work done by the force when the mass has rotated through an angle $d\theta$ is

$$dW = \vec{F} * d\vec{s}$$
$$dW = |\vec{F}| r d\theta$$

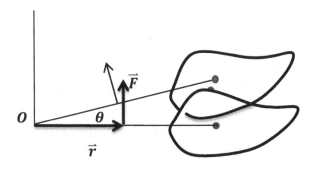

$$figure\ 4.14$$

Kinetic Energy of Rotation

Note that $|\vec{F}|r$ is the torque:

$$W \quad = \quad \int |\vec{\tau}| d\theta$$

This is the work integral expressed in terms of the angular parameters.

The result will follow from the work-energy relation:

Work   =   Change in Kineti Energy

$$\int \vec{F} * d\vec{s} \quad = \quad \frac{1}{2} m(v_f^2 - v_i^2)$$

$$\int |\vec{\tau}| d\theta \quad = \quad \frac{1}{2} mr^2(\omega_f^2 - \omega_i^2)$$

$$\Rightarrow \quad \int |\vec{\tau}| d\theta \quad = \quad \frac{1}{2} I (\omega_f^2 - \omega_i^2)$$

This is the result for a point mass. As before, the mass can be considered as a differential piece of an extended body (see **section 4.3.1**). These can now be added (integrated) for all the points in the extended body: the torque integrates to the net external torque applied; the moments of inertia integrate to the moment of inertia of the extended body. Therefore, the last equation above stays intact throughout the integration process and remains valid in the above form.

The result could also have been obtained from the torque equation for an extended body:

$$\vec{\tau} \; = \; \frac{d\vec{L}}{dt}.$$

Multiply both sides by $d\theta$ and integrate:

$$\int |\vec{\tau}|\, d\theta \;\; = \;\; \int \frac{d(I\omega)}{dt}\, d\theta$$

$$\int |\vec{\tau}|\, d\theta \;\; = \;\; \int \frac{d\theta}{dt}\, d(I\omega)$$

$$\int |\vec{\tau}|\, d\theta \;\; = \;\; I \int \omega\, d(\omega)$$

$$\int |\vec{\tau}|\, d\theta \;\; = \;\; \frac{1}{2}I(\omega_f^2 \, - \, \omega_i^2)$$

The last expression is the analogue of the work-energy relation. In words, it is

> *the work delivered by the torque equals the change in the kinetic energy of rotation.*

4.6 THE ANALOGY BETWEEN LINEAR AND CIRCULAR PARAMETERS

4.6.1 Tabulation of Results

All of the basic quantities of angular motion have been defined. The similarity of the torque equation to the dynamics equation has resulted in laws of angular motion which are the analogues of those which apply to linear motion. The following is a tabulation showing the correspondence.

ANALOGOUS QUANTITIES

LINEAR		ANGULAR	
Name	Symbol	Name	Symbol
position	$\vec{s}$	*ang pos*	Θ
Velocity	$\vec{v}$	*ang vel*	Ω
acceleration	$\vec{a}$	*ang acc*	A
Mass	*m, M*	*mom of Inert*	I
momentum	$\vec{p}$	*ang mom*	$I\vec{\omega}, \vec{L}$
impulse	$\vec{I}, \vec{M}, \int \vec{F}\, dt$	*ang Imp*	$j, \int \vec{\tau} dt$
Work	$W, \int \vec{F} * d\vec{s}$	*work by trq*	$\int I d\theta$
kinetic enrgy	$T, \frac{1}{2}m v^2$	*k.e. of rot*	$\frac{1}{2}I\omega^2$
Potent enrgy	$V, \frac{1}{2}kx^2$	---	---

figure 4.15

Analogy between Linear and Angular Parameters

BASIC LAWS

impulse   = Δ *momentum*

angular impulse   = Δ *angular momentum*

work   = Δ *kinetic energy*

work delivered by torque   = Δ *k.e. of rotation*

all forces conservative $\Rightarrow$ *energy is conserved*

H = $T + V$ =   *constant*

The study of this table is not only a memory aid to help keep track of the many quantities but it also serves to help in the understanding of the similarities in the roles played by these quantities.

4.7 THE GENERAL ACCELERATION OF A POINT MASS
4.7.1 Description of the System

To avoid unnecessary excursions into vector-matrix algebra, attention has been restricted to motion in the *xy*-plane. We are not entertaining the thought of consorting with forces that alter the plane of motion or, equivalently, the direction of the angular parameter vectors. It is specifically the analyses of these forces that require the mathematics of vector-matrix algebra, a topic which belongs to a course in the equations of mathematical physics.

By virtue of the restriction to motions in the *xy*-plane, the vectors associated with positions, velocities and accelerations are in this plane and the vectors associated with the angular parameters point in the *z*-direction. This particularization of the angular parameters results in several rules of thumb which are **peculiar to the restrictions we have imposed.** These are the ollowing:

1. *sums and differences of angular vectors are reduced to algebraic sums and differences of their magnitudes and the resultant is in the z-direction. (Secret use of this fact has already been slipped by your scrutiny in section 4.5.3.);*

293

2. *the dot product of two angular vectors is the product of their magnitudes; the cross-product of two angular vectors is zero;*

3. *the dot product of an angular vector with a vector in the xy-plane is zero; the magnitude of the cross-product of an angular vector with a vector in the xy-plane is the product of their magnitudes; the direction of such a cross-product is in the xy-plane and normal in the right-hand rule sense to the vector in the xy-plane which appears as one of the factors in the product.*

When vectors were expressed in terms of two stationary orthogonal Cartesian coordinates (like x and y) things were much neater: what happened in the x-direction was its own business and was related only to the x-components of forces, velocities, accelerations, etc.; and similarly, for the y-direction. This followed from the fact that the x- and y-coördinates (and hence their directions and associated unit vectors) were fixed in space.

It is generally the case that the individual directions of non-fixed coördinate systems do **not** mind their own business: there is cross-talk between them. In other words, what happens in one direction affects what goes on in the others. Note that what is being addressed here occurs because the coördinate systems are subject to changes as the motion progresses; additional cross-talk between the different directions occurs also when the coördinate directions are not normal to each other and/or when curved coördinate systems are used. There will be no occasion here to deal with either of these two cases.

It will be necessary however, to deal with cylindrical coordinates (radial, angular and **z**-directions). In cylindrical coördinates, only the **z**-direction is fixed: the radial and angular directions vary as the mass changes position and consequently the quantities associated with angular motion are affected.

In **figure 4.16**, a point mass is shown moving in the **xy**-plane along an otherwise arbitrary path: all parameters of the motion and their derivatives are arbitrary. The problem is to find the acceleration and hence the force necessary to produce the motion. We suppose therefore, that the position of the mass is known at every instant.

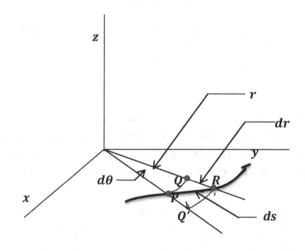

figure 4.16

Accelerations in the xy-plane

4.7.2 Explanation of the Diagram

The mass is shown at the two points P and R differentially displaced from each other by a distance ds. Suppose that the mass is at P at time t and at R at time $(t + dt)$. The problem is to find the force $\vec{F}$ acting on the body at time t which will get it from P to R in time dt. Since $\vec{F}$ and $\vec{a}$ are proportional via the dynamics equation, the problem is equivalent to the kinematics problem of determining the acceleration the body undergoes during that time: once the acceleration is known, the force is known.

We will express all quantities in the analysis in terms of the radial and angular components of the vectors involved.

To this end, we define a differential sized vector $\vec{ds}$ which goes from P to R and write it as the sum of its components in the radial and angular directions:

$$\vec{ds} \;=\; dr\,\vec{\iota_r} + rd\theta\,\vec{\iota_\theta}.$$

In this expression, the first term on the right represents the line from Q to R and the second represents the line from P to Q. The vectors $\vec{\iota_r}$ and $\vec{\iota_\theta}$ are unit vectors in the radial and angular directions, respectively. They are defined similarly to the unit vectors $\vec{\iota_x}, \vec{\iota_y}$ and $\vec{\iota_z}$ which were used for the representation of vectors in a Cartesian coördinate system: $\vec{\iota_r}$ is a unit vector pointing in the radial direction (from the origin toward the mass whose motion is being described); $\vec{\iota_\theta}$ is a unit vector perpendicular to it pointing in the direction of increasing θ.

'Divide' this expression by dt and it follows that the velocity from P to R is

$$\vec{v} \;=\; \frac{\vec{ds}}{dt} \;=\; \frac{dr}{dt}\vec{\iota_r} + r\frac{d\theta}{dt}\vec{\iota_\theta},$$

$$\vec{v} = \frac{d\vec{s}}{dt} = \frac{dr}{dt}\vec{\iota_r} + r\omega\,\vec{\iota_\theta}.$$

The first of the terms on the right is the rate of change of distance from the origin which is the radial component of the velocity v_r and the second is the tangential component of the velocity v_θ.

The acceleration required to get from P to R is the time derivative of this expression:

$$\vec{a} = \frac{d\vec{v}}{dt}.$$

A problem arises at this point. The unit vectors defined above for the radial and angular directions are not constant. In order to find the time-derivative of the velocity vector $\vec{v}$ it is necessary to find expressions for the time-derivatives of the unit vectors $\vec{\iota_r}$ and $\vec{\iota_\theta}$. This issue is addressed in the next section.

4.7.3 The Formal Derivation of the Acceleration

If you are not comfortable with vectors yet, this is a good time to review **sections 0.2, 3.2.6** and **3.2.7** together with their subsections.

Note that in the case of Cartesian coordinates each of the major directions is freed from the other two in the sense that the force applied in a given direction affects a reaction of the system in that direction only: each of the three problems (one for each major direction) can be solved independently of the other two. In other words, each direction minds its own business.

'Then, how come when we write everything in terms of radial, angular and z components, the different directions get mixed up?', you ask. The answer to this question was already alluded to above. Here is a more detailed description.

Recall that the unit vectors $\vec{\imath}, \vec{\jmath}$ and $\vec{k}$ which are used to express vectors in Cartesian coödinates are constant vectors: each is of unit length and points in one of the three (fixed) coördinate directions. This is the essential fact that kept the directions independent of one another. The unit vectors in the (r, θ, z)-coördinates however, are not constant: $\vec{\imath}_r$, the unit vector in the radial direction points from the origin in the direction of the moving body and hence changes as the motion progresses; $\vec{\imath}_\theta$, the unit vector in the angular direction, is always normal to $\vec{\imath}_r$ and hence changes in a similar fashion. Since the unit vectors $\vec{\imath}_r$ and $\vec{\imath}_\theta$ change with time which implies they have non-zero time-derivatives. The unit vector in the z-direction $\vec{\imath}_z$, is constant and, as in the case of Cartesian coördinate systems, has a zero time-derivative.

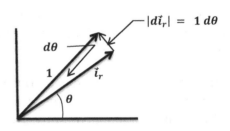

$$|d\vec{\imath}_r| = 1\, d\theta$$

figure 4.17

The Time-Derivatives of the Unit Vectors

The issue to be addressed then is that of finding the time-derivatives of the unit vectors $\vec{\imath}_r$ and $\vec{\imath}_\theta$, so that the acceleration vector can be determined as the time-derivative of the velocity vector.

Recall that $\vec{\imath}_r$ varies only in direction: its magnitude is always unity. We know a vector trick that applies: namely, any derivative of a constant magnitude vector is normal to the vector. Didn't that work out nicely! That's the direction of $\vec{\imath}_\theta$. All that remains to be done is to find its magnitude.

The change in $\vec{\imath}_r$ when the body moves through an angle $d\theta$ is (see *figure 4.17*)

$$|d\vec{\imath}_r| \quad = \quad |\vec{\imath}_r|d\theta$$

'Dividing' this equation by dt:

$$\frac{d|\vec{\imath}_r|}{dt} \quad = \quad \omega$$

This is the magnitude of the derivative. Since it points in the angular direction, it can be written

$$\frac{d\vec{\imath}_r}{dt} \quad = \quad \omega\vec{\imath}_\theta$$

Since the angular unit vector is always normal to the radial unit vector, it rotates through the same (differential) angle in the same (differential) time: therefore,

$$\frac{d\vec{\imath}_\theta}{dt} \quad = \quad -\omega\vec{\imath}_r$$

Exercise: Verify the last assertion.

4.7.3.1 The Derivation of the Acceleration Equation

Now that all the mathematical machinery is in place, it is a simple matter to find the acceleration of an arbitrary motion in the **x-y** plane in terms of the angular (polar) parameters. What follows starts over with the position vector. The first time-derivative will be the velocity vector and will be seen to be in agreement with the expression found at the end of the previous section. Let $\vec{r}, \vec{v}$ and $\vec{a}$ be the position, velocity and acceleration vectors, respectively and r the scaler distance from the origin. Then the position vector is given by

$$\vec{r} \;=\; r\vec{i}_r$$

The velocity is found by taking the time-derivative treating both factors on the right as variables:

$$\vec{v} \;=\; \frac{d\vec{r}}{dt} \;=\; \frac{dr}{dt}\vec{i}_r \;+\; r\omega\,\vec{i}_\theta$$

(Compare this to the expression at the end of the previous section.)

The acceleration is the next time-derivative:

$$\vec{a} \;=\; \frac{d\vec{v}}{dt}$$

$$\vec{a} \;=\; \frac{d}{dt}\!\left(\frac{dr}{dt}\,\vec{i}_r \;+\; r\omega\vec{i}_\theta\right)$$

$$\vec{a} = \frac{d^2r}{dt^2}\vec{i}_r \;+\; \frac{dr}{dt}\omega\,\vec{i}_\theta \;+\; \frac{dr}{dt}\omega\,\vec{i}_\theta \;+\; r\alpha\,\vec{i}_\theta \;-\; r\omega^2\,\vec{i}_r$$

When the terms are collected and the result is written in terms of its radial and angular components,

$$\vec{a} \;=\; \left(\frac{d^2r}{dt^2} \;-\; \omega^2 r\right)\vec{i}_r \;+\; \left(r\alpha \;+\; 2\omega\frac{dr}{dt}\right)\vec{i}_\theta$$

The vector formalism developed above has accomplished what we intended. It has successfully collected all the scattered facts about the equations of motion expressed in angular parameters into a simple direct theory. Note however, that while the formalism facilitates the description of physical systems, it offers little insight into the origin of the individual terms of the general acceleration equation. It is the analyses of *section 4.7.2* which foster an understanding and it is this formalism which enables easy application. Both of these ingredients are necessary to develop a good sense of the application of vectors to the description and solution of problems.

It is helpful to pause at this point to identify the individual terms in the above expression for the acceleration of a mass moving in two dimensions. To this end suppose an external force $\vec{F}$ is expressed in (r, θ)–coördinates. Then we have

$$\vec{F} = m\vec{a}$$

$$F_r \vec{\iota_r} + F_\theta \vec{\iota_\theta} = \left(m\frac{d^2r}{dt^2} - m\omega^2 r \right)\vec{\iota_r} + \left(mr\alpha + 2m\omega\frac{dr}{dt} \right)\vec{\iota_\theta}.$$

Since this is a vector equation, the corresponding components must be equal. We therefore have the two equations

$$F_r = m\frac{d^2r}{dt^2} - m\omega^2 r$$

and
$$F_\theta = mr\alpha + 2m\omega\frac{dr}{dt}$$

in the two unknowns $r(t)$ and $\theta(t)$. (The function $\theta(t)$ enters the equations via ω and α which are its first and second time-derivatives.)

Look at each term. The first term in the radial equation is the mass times the acceleration in the radial direction; this is a simple linear '*ma*' term. The second term in the radial

301

equation is the centripetal force which tends to alter the direction of motion and therefore makes it move in a curve. The first term in the angular equation is the tangential force, i.e. the force in the direction of motion that results from the angular acceleration α. The second term in the angular equation has not been seen before. Unfortunately, this force has no name although it is somewhat akin to the Coriolis force which will be discussed later (see **section 4.7.3.3**). This force will be referred to here as the 'mystery force.'

4.7.3.2 The Mystery Force

The mystery force has the form

$$F_M \;=\; 2m\omega\frac{dr}{dt}$$

where r is the scaler distance of the mass from the origin and ω is the angular velocity. This force acts in the angular direction.

To investigate the nature of this force, first note that it is non-zero only when both the angular velocity ω $(=\frac{d\theta}{dt})$ and radial velocity $\frac{dr}{dt}$ are non-zero. This prompts us to suspect that its function might be clarified easily by considering what happens when one tries to walk along a radial line on a rotating disk. This is in fact considered below and will result in a geometric derivation that duplicates the formal one given above.

The situation is depicted in **figure 4.18**. Imagine the mass moving in a circle of radius r so that its tangential velocity is ωr. In a time Δt, it would travel a distance $\omega r\Delta t$. If the mass moves to a radial distance of $(r + \Delta r)$, its tangential velocity must increase so that the mass can 'keep up' with the rotational motion: if it does not, the mass will not remain on the radius. 'Keeping up' with the rotational motion however requires a side-thrust, i.e. a force in the angular direction.

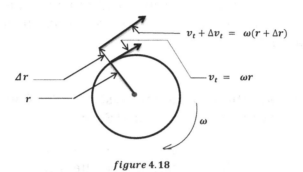

figure 4.18

The Mystery Acceleration

At the new radial distance, the mass must travel a distance $\omega(r + \Delta r)\Delta t$ which indicates an added distance Δs given by

$$\Delta s = \omega \Delta r \Delta t.$$

If the mystery acceleration during the time increment Δt is a_m, then we must have (rf. **Section 3.3.2.1, Constant Acceleration Equations**)

$$\Delta s = \frac{1}{2} a_m (\Delta t)^2$$

$$\omega \Delta r = \frac{1}{2} a_m \Delta t$$

$$\Rightarrow a_m = 2\omega \frac{dr}{dt}$$

$$\Rightarrow F_m = 2m\omega \frac{dr}{dt},$$

303

> where F_m is the mystery force. This is the
> same result as that obtained formally in the
> previous section.

Thus, when a mass moves along a radial line in a rotating coördinate system the mystery force provides the side-thrust necessary to adjust its speed to match the angular velocity at the new radius. Without this force, the mass will not stay on the radial line.

4.7.3.3   Rotating Coördinate Systems: The Coriolis Force; The Centrifugal Force

The study of rotating coördinate systems will now be carried a little further because it offers insight into an interesting ilk of phenomena. For a moment consider that a planet is a rotating system. The phenomena referred to are things such as hurricanes, tornados, whirlpools, typhoons, dirt devils, etc. They are all similar: in general they are circulatory motions of gaseous or liquid masses on the surface of the planet.

It is the rotation of the planet around its axis that tends to induce these motions at its surface. It should be pointed out however that in many cases this tendency is so miniscule that it is easily eradicated by other factors which may be present. For example, the angular velocity of the earth's rotation is approximately 0.000073 rad/sec so that it probably has little to do with an effect so small as the whirlpool in your kitchen sink drain.

Things such as this will be neglected here however because the current purpose is to investigate the origin of this tendency and its relation to the rotation of the planet. These things are independent of the effectiveness of the forces in play

Without further ado, consider two coördinate systems: one inertial and another rotating with a constant angular velocity $\vec{\omega}$.

The analysis will be approached as follows. An arbitrary vector will be considered and its time-derivatives as they appear in the two systems will be compared. Once the rule for this comparison is determined, it will be applied to the position vector. For the first time-derivative, the velocity as seen in the two systems will be related; for the second time-derivative the acceleration as seen in the two systems will be related.

Let $\vec{Q}$ be some vector whose time derivatives in the inertial and rotating systems are to be compared.

Note first that the inertial system is the home of Newtonian mechanics and the time derivative taken in this system is simply the ordinary change in $\vec{Q}$ per unit time. This derivative will be designated[17]

$$\left(\frac{d\vec{Q}}{dt}\right)_i$$

where the subscript indicates that it is the rate of change of $\vec{Q}$ as seen in the inertial system. The derivative taken in the rotating system will be designated

$$\left(\frac{d\vec{Q}}{dt}\right)_r$$

where the subscript indicates that it is the rate of change as seen in the rotating system.

---

[17] The subscript 'i' will refer to the inertial system; the subscript 'r' to the rotating system.

305

The idea is simple. We assert that changes in $\vec{Q}$ as seen in the inertial system equal the changes as seen in the rotating system plus those changes which are induced by the rotation:

$$\left(\frac{d\vec{Q}}{dt}\right)_i \;=\; \left(\frac{d\vec{Q}}{dt}\right)_r \;+\; \vec{\omega} \times \vec{Q}.$$

This is the relation that allows us to compare time-derivatives in the two systems.

Recall that $\vec{Q}$ was an arbitrary vector. We now identify $\vec{Q}$ with the position vector $\vec{r}$. The term on the left becomes the time-derivative of the position as seen in the inertial system which is the velocity in the inertial system, $\vec{v}_i$. Continuing in thgis way, the entire relation becomes

$$\vec{v}_i \;=\; \vec{v}_r \;+\; \vec{\omega} \times \vec{r}.$$

The accelerations (and hence, the dynamics in the two systems) can now be related by taking the time-derivative of this last expression:

$$\left(\frac{dv_i}{dt}\right)_i = \left(\frac{d\vec{v}_r}{dt}\right)_r + \vec{\omega} \times \left(\frac{d\vec{r}}{dt}\right)_r + \vec{\omega} \times \vec{v}_r + \vec{\omega} \times (\vec{\omega} \times \vec{r})$$

$$\vec{a}_i \;=\; \vec{a}_r \;+\; 2\vec{\omega} \times \vec{v}_r \;+\; \vec{\omega} \times (\vec{\omega} \times \vec{r})$$

$$\Rightarrow m\,\vec{a}_r \;=\; m\,\vec{a}_i \;-\; 2m\vec{\omega} \times \vec{v}_r \;-\; m\,\vec{\omega} \times (\vec{\omega} \times \vec{r}).$$

The last relation has been multiplied through by a mass m so that each term represents a force on, rather than an acceleration of m. It will presently be used to determine how the situation looks from the rotating system, i.e. from the planet surface.

Identify each of the terms in the equation:

1. $m\vec{a}_r$ is the force on the mass as seen in the rotating system. It is a sum of real and fictitious forces. Designate it $\vec{F}_r$;
2. $m\vec{a}_i$ is the force on the mass as seen in the inertial system. This is the only real force in the equation. Designate it $\vec{F}_i$;
3. $-2m\vec{\omega} \times \vec{v}_r$ is called the Coriolis force. It is a fictitious force. Its presence is an illusion brought about by the rotation of the system. Note its similarity to the mystery force discussed earlier. It will be designated $\vec{F}_C$;
4. $-m\vec{\omega} \times (\vec{\omega} \times r)$ is called the centrifugal (center-fleeing) force. It too is a fictitious force resulting from the rotation of the system. It will be designated $\vec{F}_{cf}$.

By jumping into the accelerated (rotating) system, we have pre-empted the section on D'Alembert's Principle whose business is the analyses of such systems. Of central importance in this context is the fact that

> *when viewed with respect to an accelerated coödinate system, all objects behave as though the fictitious forces that accompany the acceleration were really present. (See sections 5.5.1 and 5.5.2.)*

The force equation derived above is the central relation used in this context. Keep in mind that the current investigation allows rotating systems to be related to inertial systems when dealing with issues related to Coriolis forces or D'Alembert's Principle.

4.7.3.4   Application to a Thought Problem

Problem

Imagine a frictionless horizontal disk rotating with constant angular velocity ω counter-clockwise around the vertical axis through its center. Suppose now that a mass m is placed on the disk at a distance r from its center.

The first part of the problem is to describe the motion of the mass as seen from the inertial frame of reference and also as seen from a coördinate system that rotates with the disk. This is fairly easy and can be done by inspection.

The second part of the problem is to reconcile this description with the equations and principles of physics which constitute the mathematical description of the system. This second part can become elusive and intricate and has resulted in the gnashing of many a tooth (my own included). It will be seen that the analysis presented above offers a clear and tractable approach to this problem.

Solution

Motion of the mass (by inspection)

a. **From an Inertial frame**

Since the disk is frictionless, the mass simply stays where it was placed without moving while the disk slides beneath it.

b. **From the Frame Rotating with the Disk**

Imagine yourself at the center of the disk and rotating with it. From your viewpoint the mass rotates clockwise in a circular path whose radius is its initial distance from the center.

Mathematical Description

The plot afoot is this: start with the position vector and use the result of the previous section to relate its second time-derivatives, i.e. its accelerations, in the two systems. In this way, the apparent forces acting on the mass from the two viewpoints will be related.[18] It will then be shown that

(no net force in the inertial system) $\Rightarrow$

(a centripetal force $m\omega^2 r$ in the rotating system)

and the motion of **m** from both viewpoints will be reconciled with the mathematical description.

We have

$$\vec{r} \;=\; r\vec{\imath}_r$$

where r is the scaler distance of **m** from the axis of rotation. Then

$$\left(\frac{d\vec{r}}{dt}\right)_i \;=\; \left(\frac{d\vec{r}}{dt}\right)_r \;+\; r\vec{\omega}\times\vec{\imath}_r$$

$$\vec{v}_i \;=\; \vec{v}_r \;+\; r|\vec{\omega}|\,\vec{\imath}_\theta$$

$$0 \;=\; \vec{v}_r \;+\; r|\vec{\omega}|\,\vec{\imath}_\theta$$

$$\Rightarrow\quad \vec{v}_r \;=\; -r|\vec{\omega}|\,\vec{\imath}_\theta.$$

---

[18] There is no real net force acting on m. As is always the case with accelerated coördinate systems, fictitious forces arise and from the viewpoint of the accelerated system, things behave as though they are real. See section 5.5.1.

The result of taking the next time-derivative and using this last relation is

$$0 \; = \; \left(\frac{d\vec{v}_r}{dt}\right)_r + \; r|\vec{\omega}| \left(\frac{d\vec{\iota}_\theta}{dt}\right)_r + \; r|\vec{\omega}|^2\vec{\iota}_r \; - \; r|\vec{\omega}|^2\vec{\iota}_r$$

$$0 \; = \; \vec{a}_r \; + \; 2\,r|\vec{\omega}|^2\vec{\iota}_r \; - \; r|\vec{\omega}|^2\vec{\iota}_r$$

$$\Rightarrow \vec{F}_r \; = \; m\vec{a}_r \; = \; -m\,r|\vec{\omega}|^2\vec{\iota}_r\,.$$

The last relation indicates that the net radial force is equal to the centripetal force necessary to cause the mass to move in a circle of radius *r*. Keep in mind that this force is the sum of two fictitious forces (the Coriolis and the centrifugal). It follows that it too is fictitious. But don't get weird about what that means because the corresponding circular motion of the mass is illusionary. Both are the result of viewing the situation from an accelerated coördinate system. In other words

> *the fictitious force and the illusionary motion it causes are a set which enter and leave your considerations together as you enter and leave the rotating system.*

Incidentally the expression obtained above for the velocity $\vec{v}_r$ as seen from the rotating system indicates that the apparent motion of *m* is in the clockwise direction.

Hence the motion and its mathematical description are reconciled.

Note that this problem could have been solved by the direct application of the equation derived in the previous section in which the four forces involved were related. Starting with this equation obtain the same result as that shown above.

4.8   SUMMARY AND A FEW REMARKS BEYOND

The point was made in this chapter that the equations of motion can be expressed in many different ways and that t

+he complexity of the equations can vary greatly depending on the choice of parameters.

The importance of circular motion and the motions in centrally directed force fields motivated the definition of the angular parameters of motion and an extensive investigation of their properties. (A limited version of) the dynamics equation was then expressed in terms of these parameters resulting in the torque equation whose similarity to

$$\vec{F} \;=\; \frac{d\vec{p}}{dt}$$

prompted the definitions of angular analogies to some of the linear quantities, namely angular impulse, angular momentum, work done by a torque and kinetic energy of rotation. It was illustrated that their roles in angular motion were analogous to those of their linear motion analogues. In particular, the angular laws, **angular impulse = change in angular momentum** and **work done by a torque = the change in kinetic energy of rotation** were derived from the torque equation in a way similar to that in which their linear analogues were derived from the dynamics equation.

The quantities and equations which pertain to the circular motion of a system around an axis are basically more complex than those associated with linear motion. An example is the moment of inertia which accounts for the inertial effects of the mass when the motion is rotational. In linear motion, the mass is a scaler property of the body which quantifies its resistance to a change in motion; in rotational motion, its analogue, the moment of inertia, quantifies the body's resistance to a change in rotational motion, but the moment of inertia depends on the axis of rotation and the mass distribution around it; it also involves the size and shape of the body.

This situation is further complicated by the following. In linear motion, the momentum $\vec{p} = m\vec{v}$ is such that in all contexts, it points in the same direction as the velocity and the mass m thus retains its status as a simple scaler. The angular momentum has an analogous definition $\vec{L} = I\vec{\omega}$, but in the general case $\vec{L}$ and $\vec{\omega}$ do not point in the same direction. Hence, the moment of inertia cannot be a simple scaler. However, it still represents the body's resistance to a change in rotational motion which provides considerable motivation to retain the definition of $\vec{L}$, but to do so requires the redefinition of the moment of inertia I, as a 3×3 matrix which adjusts both the magnitude and direction of $\vec{\omega}$ to satisfy its role in the angular momentum definition. This problem was avoided by restricting the motions considered here to two dimensions.

In general, the consideration of motion around an axis adjoins the intricacies of space geometry to the dynamics of motion. Each of the many kinematic relations involved was found to have its origin in $\vec{F} = m\vec{a}$, but the relations emerged from the analyses as isolated facts. The use of the basic properties of the unit vectors $\vec{\imath_r}$ and $\vec{\imath_\theta}$ allowed for a formal description that organized the results into a more coherent whole.

Some special attention was given to the centripetal acceleration (force) which quantifies the change in direction of motion along a curved path, and to the Coriolis acceleration (force) which quantifies the side-thrust required to maintain a given angular velocity as you move along a radial line.

Throughout the chapter, focus has been shifted away from the natural decomposition of vectors into components tangential and normal to the path of motion and toward their decomposition into radial and angular components. In the case of pure circular motion, these viewpoints are identical.

The question arises, 'why relinquish a description which proceeds so naturally from the dynamics equation?'. The answer is that it leads to too many mathematical complications. They arise from the fact that such a description calls for knowledge of the position and radius of the circle which best approximates the path of motion at the current position of the mass. These are called the center and radius of gyration and are related to the path of motion in a non-linear fashion. Since the path is unknown until after the problem is solved, these quantities enter the equations of motion as unknowns. The crippling factor to this approach is that the resulting non-linear differential equations of motion are too difficult to solve. The unfortunate fact is that this description which potentially offered by virtue of its relation to the general equations of dynamics, a valuable approach to many mechanical problems is disqualified as a viable path to their solution.

With the completion of this survey of the basic description of motion around an axis, the path is open to analyze a number of diverse mechanical systems. The principles which are the subject of this chapter will accrue content and afford deeper insights when their roles in the analyses of specific problems are seen. This is the subject of *Chapter V*.

Chapter IV: The Angular Parameters

CHAPTER V

Applications of the Angular Parameters

5.1 FORMAT AND SUGGESTIONS

5.1.1 Peripheral Developments: Filling in the Blanks

This chapter contains a number of sample problems with their solutions interspersed with developments, explanations and/or refinements of the material in **Chapter IV.** Each problem is accompanied by a verbal analysis which outlines the principles which apply and organizes a 'path' to the solution. In a sense, this is the most elusive part of solving a problem when you are new at it. It requires a 'sense' of the entire problem and how it relates to the principles you have been studying. The actual calculations are a way of translating this 'gestalt' insight into a sequential process.

The development of this sense is the essence of becoming a master in any field of endeavor and the study of mechanics takes no exception. You are urged to focus attention on this verbal analysis and to develop your own ability to 'see' the solution before you embark on the determination of the details. In some cases this may not be immediately possible and floundering with whatever unorganized facts you can gather may well serve the development of your 'seeing' ability. When this is the case, take a moment after your fact-gathering to abstract from your information an overall view of the problem. It will be well worth your while.

In keeping with these remarks, it is of value to read the problem and attempt your own analysis in general terms before reading the verbal analysis presented and this can be supplemented with some actual calculations whenever this is feasible. It is more often the case than not that there are two or more ways to approach a given problem. If your way differs from the one presented, follow it through. It may very well be equally as valid.

There is some new material in this chapter: the equations of statics (the study of stationary systems like buildings and bridges which tend to upset us when they are not stationary); rolling friction; D'Alembert's principle; power.

STATICS is the branch of mechanics that deals with structural stability: it is the analysis of systems which are in a state of static equilibrium. There is nothing really new here except the experience of actually doing specific problems. The requirements are that the net force on, and the net torque around any point of the system is zero. If either of these is non-zero, something will move and the system will no longer be static.

ROLLING FRICTION is a friction force that places a constraint on two surfaces. Without it, a cylinder will slide instead of roll down an inclined plane. (see *section 3.1.3* for a discussion of this.) When there is rolling friction, the surface of the cylinder is always in non-slipping contact with the surface of the plane. This establishes a relation between the angular velocity around its axis and its linear velocity down the plane: this relationship is the equation of constraint. Two important features of rolling friction are that it does not dissipate any energy as heat and it appropriates the energy between the rotational and translational motion in exactly the right proportions to maintain non-slipping contact.

D'ALEMBERT'S PRINCIPLE is a mathematical triviality but in terms of the physics involved, it represents a dramatic shift in viewpoint. Instead of considering yourself as comfortably at rest in an inertial frame of reference watching some system perform all its terrible gyrations, you imagine yourself *at rest with respect to the accelerated coördinate system.* When you do this, the law of inertia and the law of dynamics no longer hold from your viewpoint. D'Alembert's principle determines the adjustments to these laws that must be made to accommodate your non-inertial viewpoint.

POWER is a very specific quantity in mechanics. It is defined as the rate at which work is delivered to or by a system. When the system is conservative, it corresponds to the rate of increase of kinetic energy. The issue when considering power is *how fast* the work is done: if a machine can deliver a certain amount of work to a system in a time *t*, another machine with half the power can deliver the same amount of work but it takes twice as long.

Once the principles of translational and rotational motions are understood, you have the background to analyze quite a large repertoire of mechanical systems. A first course in mechanics usually focuses attention on the nature and applications of these principles; a second course in mechanics is concerned primarily with the mathematical methods which were devised to analyze more complex systems. The point is that now is the time to get a solid understanding of the principles. If you pursue this subject further you will need to bring this understanding with you because the emphasis will be on the mathematical methods of their application.

This chapter is designed to clarify the application of the principles which have already been presented and to complete the survey of classical mechanics by presenting those not yet covered. The actual sample problems were chosen to indicate the wide variety of systems which can be analyzed using the material presented prior to this point. Use this chapter to learn to think as a physicist, i.e. to look at systems in terms of the basic mechanical laws and make general qualitative judgments to guide you through the details.

5.2   STATICS

5.2.1   The Equations of Statics and Their Solution

The equations of statics are simple:

$$\sum \vec{F} = \vec{0}$$

$$\sum \vec{\tau} = \vec{0}$$

These conditions must hold for the system as a whole as well as for every component of the system. This is obvious because the condition that a system be static is that it and all its component parts have zero linear and angular accelerations: the above conditions are the linear and angular dynamics equations with the linear and angular accelerations set to zero.

In the application of these conditions, one chooses points of interest in the system. For example, consider a rope under tension: the points of interest are where the rope interacts with other parts of the system, namely, at its endpoints or where it goes over a pulley. In contrast, one could apply the conditions to a differential piece of the rope (assumed massless) but that would not help solve the problem: the tension on one side of the piece would be equal and opposite the tension on the other: thus the piece of rope does not accelerate and it is seen that the tension is simply transmitted along the rope to a point of interaction with another part of the system.

Consequently, one generally applies the equations at those points where one component of the system joins with another. The method is to draw a diagram showing all the forces, known and unknown, and write out the conditions that they add to zero. In this way it is possible to gather enough independent equations to find all the unknown forces. The equations of statics are generally linear, so the mathematics involved in the solution to a statics problem is that of solving a set of simultaneous linear equations.

Each component of the system is treated like a given unit that interacts with the others at its points of contact. When one relaxes the attitude that some component is a given unit and wants instead to find its state in the context of the system, then he must consider the details of the forces internal to that component. An example would be the deformation of a structural beam under a certain load and subject to a certain compression. In this case the beam is no longer a 'given'. One would first find the external forces acting on the beam (contact forces, gravity, load, etc.) and then consider their effects on the internal points of the beam to find its deformation.

This example illustrates the conditions under which the attitude that the forces and torques need be considered only at the points of contact breaks down. For our purposes here however, each component will be considered as a given unit except in the case of sample problem 3 which is included as an example of a statics problem whose solution requires a differential analysis.

5.2.2 Sample Problems in Statics

Sample Problem 1: A Pivoted Boom

A beam of length L has one end attached at P to wall by a pin and the other attached by means of a wire to the same wall directly above the pin so that the beam is horizontal as shown in *figure* 5.1.*a*. A mass m is suspended from the beam at a distance l from the wall. Find the tension T in the wire and the shear force S at the pin in terms of the distances, the mass and the angle between the wire and the beam.

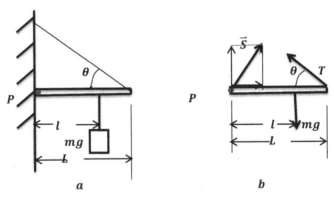

figure 5.1

A Pivoted Boom

There are three unknowns to be determined: the magnitude of the tension (its direction is known); the magnitude and direction of the shear force (or equivalently, its *x*- and *y*-components). Therefore, three equations are needed. We shall use the facts that the *x*-components of force add to zero; the *y*-components of force add to zero and the torques around the pin add to zero. That will give us three independent equations in three unknowns.

$$(\textstyle\sum F_x \;=\; 0): \quad S_x - T\cos(\theta) \;=\; 0 \tag{1}$$

$$(\textstyle\sum F_y \;=\; 0): \quad S_y + T\sin(\theta) - mg \;=\; 0 \tag{2}$$

$$(\textstyle\sum \tau_p \;=\; 0): \quad LT\sin(\theta) - mgl \;=\; 0 \tag{3}$$

Solution:

The third equation can be solved for *T*:

$$T \;=\; \frac{mgl}{L \sin(\theta)}$$

The two components of *S* can now be found by substituting this value of *T* in equations **(1)** and **(2)**:

$$S_x \;=\; \frac{mgl\cot(\theta)}{L}$$

$$S_y \;=\; mg\left(1 - \frac{l}{L}\right)$$

(If torques were taken around any point other than *P*, all three unknowns would have appeared in the torque equation and the algebra would have been more cumbersome but not incorrect.)

Discussion:

If you were designing such a system, the tension *T* would tell you how strong the wire must be to support the mass; similarly, the shear force that the pin must withstand is determined by the magnitude of the shear force $\vec{S}$. Using Pythagoras' theorem on the components of $\vec{S}$ and some trig,

$$|\vec{S}| = mg\sqrt{\left(\frac{l}{L}\right)^2 csc^2(\theta) - 2\frac{l}{L} + 1}$$

Since we left the given quantities unspecified, it is now possible to see if the answer makes sense in some extreme cases.

case 1: $\theta = 0$

In this case the wire is along the beam and the system is useless as far as supporting the mass is concerned. The equation for the tension produces $T = \infty$. See if you can discuss this further in terms of the components of the shear force $\vec{S}$. (Hint: let the angle *(θ)* be slightly larger than zero.);

case 2: $\theta = 90°$

If θ is **90°** then the beam is hanging from the ceiling supported by the wire on one end and the pin on the other. In this case, the tension $\vec{T}$ plus the *y*-component of $\vec{S}$ should equal the weight and the *x*-component of $\vec{S}$ should equal zero;

case 3: $\theta = 90°$ and $l = L$

In this case, the mass is hanging from the ceiling by the wire. $\vec{T}$ should equal the weight of the mass and $\vec{S}$ should equal zero.

Exercise:

As a partial check, ascertain that the results above are correct in the three extreme cases described.

Sample Problem 2: Winch and Ratchet Lock

A mass m is suspended by two cables, one of which is secured to the ceiling and the other of which goes over a pulley to the wheel of a winch as shown in *figure* 5.2. When the system is locked into position by the ratchet lock on the winch, the two cables are seen to make angles of θ and φ with the horizontal. If R is the radius of the winch and r is the radius of the ratchet lock, what is the force $\vec{S}$ that must be supported by the lock? (Assume that $\vec{S}$ points in the angular direction.)

Analysis

Let T_3 be the tension in the cable from the pulley to the winch; T_1, the tension from the pulley to the point O where three cables join; T_2, the tension from the point O to the ceiling. At the winch, the net torque around the wheel must be zero. Hence, the torques produced by S and T_3 must be equal and opposite. The pulley cannot sustain any net torque (otherwise it will turn); therefore the tensions on either side are equal and opposite. The bulk of the problem centers itself at point O; the sum of the forces in the x-

direction must be zero and the same for the forces in the **y**-direction.

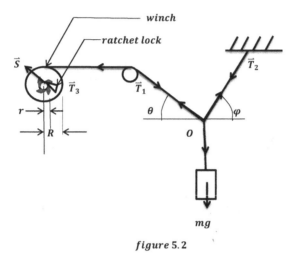

figure 5.2

Winch and Ratchet Lock

There are four unknowns T_1, T_2, T_3 and **S**. The four equations we have are: the torque equation at the winch; the equality of tensions across the pulley: the **x**-components of force at **O**; the **y**-component of force at **O**. That's four equations in four unknowns; enough to solve the problem.

Solution

The four statics equations are:

$\left(At\ winch\quad \sum \tau = 0\right)$:

$$Sr \ - \ T_3 R \ = \ 0 \qquad\qquad (1)$$

(*At the pulley*):

$$T_3 = T_1 \tag{2}$$

$$\left(At\ point\ O, \sum F_x = 0\right):$$

$$-T_1 \cos(\theta) + T_2 \cos(\varphi) = 0 \tag{3}$$

$$\left(At\ point\ O, \sum F_y = 0\right):$$

$$T_1 \sin(\theta) + T_2 \sin(\varphi) - mg = 0 \tag{4}$$

At this point the physics is done. What remains is to solve the set of 4 simultaneous equations. Proceed as follows.

Use equation (2) to eliminate T_3

$$Sr = T_1 R \tag{1'}$$

$$-T_1 \cos(\theta) + T_2 \cos(\varphi) = 0 \tag{3'}$$

$$\sin(\theta) + T_2 \sin(\varphi) = mg \tag{4'}$$

Solve equation (4') for T_2 and substitute the result into equation (3'). Solve for T_1:

$$T_1 = \frac{mg}{\cos(\theta)\tan(\varphi) + \sin(\theta)}$$

Finally, from equation (1'),

$$S = \frac{mgR}{r[\cos(\theta)\tan(\varphi) + \sin(\theta)]}$$

Exercise

Fill in the missing algebra steps. Also find T_2. You should get

$$T_2 = \frac{mg}{\cos(\varphi)\tan(\theta) + \sin(\varphi)}$$

Discussion

As a partial check of the results, the tensions will be considered in two extreme cases: when $\theta = 90°$ and when $\varphi = 90°$.

case 1: $\theta = 90°$

In this case, the mass is hanging from the pulley. Therefore, the tension T_1 should be mg and the tension T_2 should be zero;

$$T_1 \;=\; \frac{mg}{cos(\theta)\,tan(\varphi) + sin(\theta)}$$

$$T_1 \;=\; \frac{mg}{0 \cdot tan(\varphi) + 1}$$

$$\Rightarrow \qquad T_1 \;=\; mg$$

$$T_2 \;=\; y\,\frac{mg}{cos(\varphi)\,tan(\theta) + sin(\varphi)}$$

$$T_2 \;=\; \frac{mg}{cos(\varphi)\cdot\infty + 1}$$

$$\Rightarrow \qquad T_2 \;=\; 0.$$

Hence, the formulas verify the results for case 1.

case 2: $\varphi = 90°$

Comparison of the formulas for T_1 and T_2 shows that they are the same except that θ and φ are interchanged. It follows that when $\varphi = 90°$, the evaluations of T_1 and T_2 are reversed. Hence, the formulas also verify the results in case 2.

In general, simple problems in statics follow a form similar to that of the preceding two examples: set the sum of the *x*-components of force equal to zero and the same for the *y*-components and torques around any point. Eventually, you will end up with a set of simultaneous linear equations which can be solved by any of the methods available.

5.2.3   The Shape of the Main Cable of a Suspension Bridge

5.2.3.1   Description of the Problem and the Statics Equations

We turn now to a more elaborate statics problem: we will determine the (approximate) shape of the suspension cable of a bridge. In this problem, the differential equation whose solution is the shape we are after is set up using a differential analysis. To solve this equation would take us too far afield of the subject matter addressed here. Instead, the equation will be replaced by an approximate version of itself, and the replacement equation will be solved.

The aim here is to demonstrate that the simple rule of setting the net force on a differential piece of the cable equal to zero is sufficient to formulate this problem. The example also serves to illustrate not only the method by which this rule is applied differentially, but also to introduce a few often used mathematical tricks which assist in this kind of problem.

The general shape of a suspension bridge is shown in *figure* 5.3: the roadway is supported from above by a number of vertical cables which are supported in turn by the main cables which run across the top. The main cables are supported by the towers which are resting on the ground. The problem here is to determine the shape of the main cables between the two towers.

We will do nothing more than require that the *x*- and *y*-components of force on a differential piece of the cable, each add to zero. In the process of doing so, we will have to deal with the angles θ and θ' which the tensions T and T' make with the horizontal, respectively, and with the distance ds along the cable. The tensions will be eliminated and the angles and distances will be expressed in terms of the geometry of the curve. (See *figure* 5.3.) The result will be the differential equation whose solution is the shape of the cable.

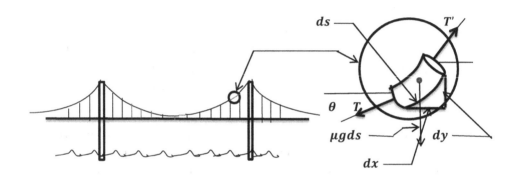

figure 5.3

Shape of the Main Cable of a Suspension Bridge

Let the origin be at the lowest point of the cable between the two towers, with x positive to the right of the origin and y, positive upward from the origin. The only forces acting in the x-direction are the x-components of the tensions:

$$\sum F_x \;\; = \;\; 0$$

$$T' \cos(\theta') \; - \; T\cos(\theta) \;\; = \;\; 0 \qquad (1)$$

Let λ be the load (weight of the road plus traffic) per unit length of roadway and μ weight per unit length of the main cable. For the y-direction, there are the y-components of the tensions, the load appropriated to a length dx of the roadway λdx, and the weight of the differential piece of cable μds:

$$\sum F_y \;=\; 0$$

$$T' \sin(\theta') - T\sin(\theta) - \lambda dx - \mu ds \;=\; 0 \qquad (2)$$

The physics is done. What remains is the mathematics necessary to eliminate the tensions and express the resulting equation in terms of x and y.

5.2.3.2 Mathematical Tricks

Recall that physicists have the (not quite legitimate) habit of treating the 'numerator' and 'denominator' of derivatives as though they were algebraic quantities. (See *section* 0.3.4.) It is seldom necessary in physics to distinguish between derivatives and the ratio of small increments.

This device can be used to express ds in terms of the geometry of the curve $y(x)$. Applying Pythagoras' theorem to the differential triangle in the diagram,

$$(ds)^2 \;=\; (dx)^2 + (dy)^2$$

$$(ds)^2 \;=\; (dx)^2 \left[1 + \left(\frac{dy}{dx}\right)^2 \right]$$

$$ds \;=\; \sqrt{1 + \left(\frac{dy}{dx}\right)^2}\; dx$$

This expression will be used to eliminate ds from the equations below.

The angles are related to the curve geometry through the first derivative:

$$\frac{dy}{dx} = tan(\theta)$$

An expression for the $cos(\theta)$ will also be needed. A handy gimmick for finding the sine and cosine of an angle when the tangent is known is illustrated in the right triangle shown below. The diagram is contrived so that A is the tangent of the angle α. We have

$$tan(\alpha) = A$$

Using the definitions of sine and cosine, it follows that

$$sin(\alpha) = \frac{A}{\sqrt{A^2 + 1}}$$

$$cos(\alpha) = \frac{1}{\sqrt{A^2 + 1}}$$

A look at the magnified section of the cable reveals that the geometry of the cable's shape is is related to the parameters through its slope,

$$tan(\theta) = \frac{dy}{dx}.$$

If the above relations are used, we have at our disposal

330

$$sin(\theta) = \frac{\frac{dy}{dx}}{\sqrt{1 + \left(\frac{dy}{dx}\right)^2}}$$

and

$$cos(\theta) = \frac{1}{\sqrt{1 + \left(\frac{dy}{dx}\right)^2}}$$

Finally, the last manipulation is this: since only a differential length is under consideration, the difference between the primed and unprimed version of any quantity is the differential of that quantity. For example, in equation (1) the difference

$$T' cos(\theta') - T cos(\theta)$$

appears. This difference is the function $T cos(\theta)$ evaluated at $(x + dx)$ minus its evaluation at x. This difference is just $d[T cos(\theta)]$. Therefore equations (1) and (2) can be written

$$d[T cos(\theta)] = 0 \qquad (1')$$

$$d[T sin(\theta)] = \mu ds + \lambda dx \qquad (2')$$

Equation (1') can be integrated:

$$T cos(\theta) = k,$$

$$\Rightarrow T = \frac{k}{cos(\theta)},$$

where k is an arbitrary constant (of integration). Substitute this last expression for T and the expression for ds found earlier into equation (2') which then becomes

$$kd\left(\frac{dy}{dx}\right) \;=\; \mu\sqrt{1 \,+\, \left(\frac{dy}{dx}\right)^2}\; dx \,+\, \lambda dx.$$

Divide by **kdx** and the final form is

$$\frac{d^2y}{dx^2} \;=\; \frac{\mu}{k}\sqrt{1 \,+\, \left(\frac{dy}{dx}\right)^2} \,+\, \frac{\lambda}{k}.$$

The function $y(x)$ that satisfies this equation is the shape of the cable.

What has been done so far is to analyze a differential piece of the main cable to find its statics equations. Subsequently, the tensions T and T' and the angles θ and θ' were eliminated to get the equations in terms of x and y. The final outcome of all this is the differential equation for the shape of the cable.

5.2.3.3 An Approximate Solution

In its present form, the last equation is difficult to solve. (Cartesian coördinates are not the best ones for this problem.) An approximate solution will be obtained by replacing the equation with one 'close' to it and solving that one. This brings up a meta-mathematical question: 'when is the solution to almost the problem almost the solution of the problem?'. There is the hidden assumption in our approach that will be sloughed over. The only answer to above question offered here is, 'sometimes'. But we're going to do it anyway.

The last equation can be made very simple by neglecting the weight of the cable. This is reasonable because the load which the cable must support (the roadway with its traffic) is much greater and hence, would be expected to have a more profound influence on the cable's shape. This simplification amounts to neglecting the second term of the differential equation. The result is

$$\frac{d^2y}{dx^2} = \frac{\lambda}{k}.$$

(Note that $\frac{d^2y}{dx^2} = \frac{d}{dx}\left(\frac{dy}{dx}\right)$. Hence the above equation implies that $d\left(\frac{dy}{dx}\right) = \frac{\lambda}{k}dx$.)

To solve this equation, integrate once with respect to x:

$$\frac{dy}{dx} = \frac{\lambda}{k}x + c,$$

where c is another arbitrary constant of integration. (The arbitrary constants that show up when you are solving a differential equation allow the solution to be 'fit' to various possible sets of initial conditions.)

Integrate once again:

$$y = \frac{\lambda}{2k}x^2 + cx + c'$$

where c' is another arbitrary constant. Except for taking the initial conditions into account, this is the final result. The form of the equation indicates that the shape of the main cable of the bridge is a parabola.[19]

To find the constants of integration from the initial conditions, proceed as follows. Recall that the origin was chosen at the lowest point of the cable (between the towers). Then when $x = 0$, $y = 0$:

---

[19] The parabola is an approximate solution. It is 'close' to the actual shape which is called a catenary. Incidentally it is the path traced out by a point on the circumference of a circle as the circle rolls along a level surface through one complete revolution.

$$y(x) \;=\; \frac{\lambda}{2k}x^2 \;+\; cx \;+\; c'$$

$$0 \;=\; 0 + 0 + c'$$

$$\Longrightarrow \qquad c' \;=\; 0$$

Also y has a minimum when $x = 0$. This implies that $\frac{dy}{dx} = 0$ when $x = 0$:

$$\frac{dy}{dx} \;=\; \frac{\lambda}{k}x \;+\; c$$

$$0 \;=\; 0 + c$$

$$\Longrightarrow \qquad c \;=\; 0$$

Since c and c' are zero, the equation for the shape of the cable is

$$y \;=\; \frac{\lambda}{2k}x^2 .$$

To find the constant k use the fact that the cable is attached to the top of the tower. Let $2L$ be the distance between the towers and H the height of the towers above the origin. Then when $x = L$, $y = H$:

$$H \;=\; \frac{\lambda}{2k}L^2$$

Solve for k and use the result in the equation for y:

$$k \;=\; \frac{\lambda}{2H}L^2 ,$$

$$\Longrightarrow \quad y \;=\; H\left(\frac{x}{L}\right)^2 .$$

he tension T which was eliminated from the equations during the above analysis is a parameter of interest to the bridge designer: it tells him how strong the cable must be to support the bridge

The tension is:

$$T \; = \; \frac{k}{\cos(\theta)} \; = \; k\sqrt{1 + \left(\frac{dy}{dx}\right)^2}$$

$$\frac{dy}{dx} \; = \; \frac{2H}{L^2}x \quad \text{and} \quad k \; = \; \frac{\lambda L^2}{2H}, \quad so$$

$$T \; = \; \frac{\lambda L^2}{2H}\sqrt{1 + \frac{4H^2}{L^4}x^2}$$

This expression indicates that the maximum tension occurs at the tower when $x \; = \; L$:

$$T_{max} \; = \; \frac{\lambda L^2}{2H}\sqrt{1 + \frac{4H^2}{L^2}}.$$

5.3   THE DYNAMICS OF CIRCULAR MOTION

5.3.1 General Remarks

Keep in mind that we are considering circular motion in the x,y-plane around some fixed point in the plane or some axis which points in the z-direction. It is often the case that the centripetal force necessary to sustain circular motion originates at the center (axis) of rotation. If there are no external forces acting on the system other than those at the axis, then angular momentum around the axis is conserved because such forces have no torque around the axis. If there are external forces which produce a net torque around the axis, then there is an angular acceleration α around the axis according to

$$\vec{\tau} \; = \; I\vec{\alpha},$$

335

where the torque and acceleration vectors point in the **z-**direction (these cases are the only ones under consideration here), and I is the moment of inertia around the axis. (Moment of inertia is always with respect to some center or axis of rotation: when it is obvious what the center of rotation is, the phrase 'around the axis' and its equivalents will be omitted.)

It is sometimes necessary to use the parallel axis theorem to find I. Below is a tabulation of the moments of inertia for several common geometric shapes around axes through their centers of mass.

Each mass moving in a circle is subject to a centripetal force of magnitude

$$F \quad = \quad m\omega^2 r$$

directed toward the center of rotation. (The mass can be differential or finite.) Recall that motion in a circle, even if the speed is constant, is accelerated motion. The centripetal acceleration is the radial component of acceleration necessary to change the direction of the velocity and the centripetal force is the associated force which produces it.

The Coriolis acceleration and force do not enter into pure circular motion. This component of acceleration is proportional to the radial component of the velocity which is zero.

SOME MOMENTS OF INERTIA

OBJECT	AXIS OF ROTATION	MOMENT OF INERTIA
thin rod	normal to rod	$\dfrac{1}{12} ml^2$
solid cube	through center of two opposite faces	$\dfrac{1}{6} ml^2$
Ring	through the *CM* normal to the plane of the ring	mr^2
hollow cylinder	coïncident with the axis of the cylinder	mr^2
solid cylinder	coïncident with the axis of the cylinder	$\dfrac{1}{2} mr^2$
hollow sphere	any axis through the center	$\dfrac{2}{3} mr^2$
solid sphere	any axis through the center	$\dfrac{2}{5} mr^2$

5.3.2 Sample Problems in Circular Motion

Sample Problem 1 Moving Mass Caught by a Pivoting Rod

One end of a thin rod of mass m and length r is attached at a point P around which it is free to rotate. The other end of the rod has a clip arrangement to catch a solid ball of mass m' and radius r' as shown in $figure$ 5.5. Assume that the action takes place on a frictionless table top. If v is the speed of the ball when it hits the clip, what is the final angular velocity of the stick-ball system? Is energy conserved? Explain.

Analysis

The initial angular moment is the moment of the linear momentum of the ball:

$$L_i \quad = \quad m'v(r + r').$$

To find the final angular momentum, it is necessary to find the moment of inertia of the system around P. Using the parallel axis theorem for the rod, its moment of inertia around one end is:

$$I \quad = \quad I_o + MR^2$$

$$I_{rod} \quad = \quad \frac{1}{12}mr^2 + m\left(\frac{r}{2}\right)^2$$

$$I_{rod} \quad = \quad \frac{1}{3}mr^2$$

Use the parallel axis theorem again to find the moment of inertia of the ball:

$$I \quad = \quad I_o + MR^2$$

$$I_{ball} \quad = \quad \frac{2}{5}m'r'^2 + m'(r + r')^2$$

The total moment of inertia after impact is the sum of these:

$$I_{total} \quad = \quad I_{rod} + I_{ball}$$

$$I_{total} \quad = \quad \frac{1}{3}mr^2 + \frac{2}{5}m'r'^2 + m'(r + r')^2$$

The final angular velocity can now be found from the conservation of angular momentum:

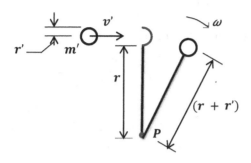

figure 5.4

Moving Mass Caught by a Pivoting Rod

$$L_f \;\; = \;\; L_i$$

$$I_{total}\omega \;\; = \;\; m'v(r+r')$$

$$\omega \;\; = \;\; \frac{15\,m'v(r+r')}{5mr^2 + 6m'r'^2 + 15\,m'(r+r')^2}\,.$$

Exercise:

Follow the analysis through once again and fill in all the missing algebra. Continue doing that in what follows until the discussion of this problem is finished.

Note: consider the limiting case when the ball is much heavier than the rod. It might be expected that the final angular velocity is the same as that of the ball just prior to impact. This is not correct however: the angular velocity just prior to impact is:

$$\omega_i \;=\; \frac{v}{r+r'} \;;$$

the final angular velocity in this case is

$$\omega_f \;=\; \frac{5m'v(r+r')}{2m'r'^2 + 5m'(r+r')^2}.$$

it is interesting to see why the judgment that the two angular velocities be the same is incorrect. The discrepancy between the two angular velocities is the result of the fact that before impact, the ball was not rotating around its own axis $[I_{ball} = m'(r+r')^2]$; after impact the ball was making one rotation per revolution around its own axis $\left[I_{ball} = m'(r+r')^2 + \frac{2}{5}m'r'^2\right]$. Since $L = I\omega$ and L is conserved, the increase in the moment of inertia I was compensated for by a decrease in the value of the angular velocity.

Finally, compare the kinetic energies before and after collision: before impact, the kinetic energy was that of the ball:

$$T_i \;=\; \frac{1}{2}m'v^2 \;;$$

after impact the kinetic energy was

$$T_f \;=\; \frac{1}{2} I_{total}\omega^2$$

$$T_f \;=\; \frac{1}{2}\frac{15[m'v(r+r')]^2}{5mr^2 + 6m'r'^2 + 15\,m'(r+r')^2}$$

Discussion

It is informative to investigate the question of lost energy a little further. To this end, consider the ratio of the final to the initial energies:

$$\frac{T_f}{T_i} = \frac{15m'(r+r')^2}{5mr^2 + 6m'r'^2 + 15m'(r+r')^2}.$$

First, notice that the numerator is always less than the denominator so that energy is always lost. What is interesting about this expression is that even when the mass of the rod is negligible (set $m = 0$), i.e. when the impact of collision is removed, the numerator is still less than the denominator and energy is still lost. Setting $m = 0$ has eliminated the effect of the inertial mass of the rod, but there is still one effect of the inelasticity of the collision which has not been eliminated: when the ball was caught by the clip, some friction was required to make the ball turn around its own axis as it revolved around P. It is this friction force which robbed the system of some of its kinetic energy.

If you are not convinced of this, note the following. The first term in the denominator of the expression for the energy ratio is the culprit: it is this term that keeps the energy ratio from equaling unity. This term came from the calculation of the moment of inertia of the ball around P. It escaped scrutiny because it was hidden in our use of the parallel axis theorem. Reference to *section 4.3.3* shows that

> *the parallel axis theorem applies only to rigid bodies.*

This is not to say that the parallel axis theorem was used incorrectly because it was not. What is being said however is that its use automatically injected into the equations the rigidity factor which in turn implied that the ball started turning around its own axis upon impact. In this way, it hid from view that it took friction to make the system rigid and fooled us into thinking that energy would be conserved when it would not.

This can be carried further. Suppose for a moment that the clip-ball contact is frictionless. Then the ball would still be constrained to move in a circle around P but its orientation would remain the same as it was before impact. In this case, the first term in the application of the parallel axis theorem would be missing because movement around its CM is missing. The inertial effects of the ball are then the same as they would be for a point mass. This is equivalent to removing the first term in the denominator of the expression for the energy ratio. The ratio would then be unity indicating that energy is conserved. The same thing happens if $r' = 0$: then the ball is a point mass and once again, energy is conserved.

Sample Problem 2: Bucket Suspended from a Windlass

A bucket of mass m' at the top of a well is attached to a rope of negligible mass to a solid cylinder of mass m and radius r (a windlass), so it can be lowered into the well as shown in *figure* 5.6. Describe the ensuing motion if the bucket is dropped from the top of the well. Given that it takes t_1 seconds for the bucket to reach the water, find the height of the well above water level. What is the velocity of the bucket at splashdown? Verify that energy is conserved between the time the bucket is dropped and the time it hits the water.

Analysis

The circular motion of the windlass and the linear motion of the bucket are related by the constraint that the length of rope that unwinds from the windlass is equal to the change in height of the bucket. The tension in the rope monitors both motions and hence adjusts itself to satisfy the constraint. The acceleration of the system can be found by eliminating the tension T from the equations of motion for the windlass and the bucket.

The acceleration will be constant and the distance and velocity equations for constant acceleration determine the height of the surface above water level and the speed at splashdown.

figure 5.5

The Windlass

This approach to the problem uses the equations of motion for a system exhibiting constant acceleration. Another approach which is somewhat simpler is to make use of the energy equations as follows.

Since the local gravity force is the only external force acting on the system, energy is conserved. If zero height is taken at water level, then all the energy is potential when the bucket is dropped and all of it is kinetic at splashdown. Hence, the initial potential energy $m'gh$ (h is the distance from water level to ground level) should equal the final kinetic energy which is the sum of the rotational kinetic energy of the windlass and the linear kinetic energy of the bucket.

Solution

Let T be the tension in the rope; a the acceleration of the bucket; α the angular acceleration of the windlass; I the moment of inertia of the windlass. For the first part of the problem, let the positive z-direction be downward with the origin at ground level. We have:

for the bucket:

$$\sum F_z = m'a;$$

$$m'g - T = m'a; \qquad (1)$$

for the windlass:

$$\sum \tau = I\alpha:$$

$$Tr = \frac{1}{2}mr^2\alpha; \qquad (2)$$

constraint:

$$a = r\alpha. \qquad (3)$$

There are three equations in three unknowns. Proceed as follows:

solve **(3)** for α and substitute into **(2)**;

$$T = \frac{1}{2}ma;$$

Substitute this value of T into (1) and solve for a;

$$a = \frac{2m'g}{m+2m'}$$

This acceleration is constant so the distance covered in a time t is

$$z \;=\; \frac{1}{2}\,at^2$$

(The last equation is the distance formula when the acceleration is constant. See **section 3.3.2.**) Therefore, the height **h** of the ground above the water is

$$h \;=\; \frac{m'g}{m + 2m'}\,t_1^2$$

The velocity of the bucket at any time is

$$v \;=\; at\,;$$

at splashdown, it is

$$v_1 \;=\; \frac{2m'g}{m + 2m'}\,t_1$$

Since the only forces acting are conservative, the previous results should verify that energy is conserved:

$$H_i \;=\; H_f$$

$$T_i \,+\, V_i \;=\; T_f \,+\, V_f$$

$$0 \,+\, m'gh \;=\; \frac{1}{2}\,m'v_1^2 \,+\, \frac{1}{2}\,I\omega^2 \,+\, 0$$

Exercise: Review this problem and fill in the missing algebra.

Exercise: Using the above results, verify that the above energy equation is satisfied.

Exercise: Solve this problem using the energy equations.

Sample Problem 3: Rotating Cylindrical Tank of Water

A closed cylindrical tank of radius r, height h and mass m is filled with water whose total mass is m'. The tank is free to rotate without friction on a vertical axis which is collinear with that of the cylinder. A small handle of negligible mass is attached to the side of the tank so that a torque can be delivered to rotate it. If an impulse K of very short duration is delivered to the tank at the handle, what is the angular velocity of the tank immediately after the impulse. What is the final angular velocity of the system? Describe the motion. What can be said about the temperature of the water?

Analysis

Since the impulse is delivered quickly, only the tank moves initially, leaving the water stationary. The tank's initial motion is governed by the angular impulse-momentum law; the moment of inertia involved in this motion is that of the tank alone. Eventually the friction between the inside of the tank and the water and the friction internal to the water itself will cause the water to move and the whole system will move as a unit. The system is free of external torques during this part of the process so angular momentum is conserved. Thus the final angular velocity can be found.

The friction forces which eventually make the tank and water move as a unit, represent a loss of mechanical energy which is transformed into heat. If the temperature increase per unit of heat energy were known, the increase in the temperature of the water during this process could be calculated.

Parallel the following analysis and fill in the missing algebra.

Solution

The moment of inertia of the tank and that of the water will be needed. To this end, start with the mass per unit area σ of the tank:

$$\sigma = \frac{m}{area\ of\ tank}$$

$$\sigma = \frac{m}{2\pi r(r + h)}$$

The moment of inertia of the tank is found by adding the moments of the top and bottom to that of the side.

The combined mass of the ends is

$$m_e = 2\pi r^2 \sigma$$

$$m_e = \frac{rm}{(r + h)}$$

This contributes an amount

$$I_e = \frac{r^3 m}{2(r + h)}$$

to the moment of inertia.

The mass of the side of the tank is

$$m_s = 2\pi r h \sigma$$

$$m_s = \frac{hm}{(r + h)}$$

which contributes an amount

$$I_s = \frac{r^2 hm}{(r+h)}$$

to the moment of inertia. (The formulas for the moments of inertia for a disk and a hollow cylinder have been used in the above.) The total moment of inertia of the tank is the sum of these:

$$I_t = I_e + I_s$$

$$I_t = \frac{r^3 m}{2(r+h)} + \frac{r^2 hm}{(r+h)}$$

$$I_t = \frac{r^2(r+2h)m}{2(r+h)}$$

The moment of inertia of the water is just that of a solid cylinder:

$$I_w = \frac{1}{2}m'r^2$$

The combined tank and water have a moment of inertia given by

$$I = \frac{r^2}{2}\left(\frac{r+2h}{r+h}m + m'\right)$$

The preliminary calculations are now completed.

At the time the impulse was delivered,

angular impulse = change in angular momentum

$$Kr = I_t \omega_i$$

$$\omega_i = \frac{rK}{I_t}$$

$$\Rightarrow \omega_i = \frac{2(r+h)K}{r(r+2h)m}$$

where ω_i is the angular velocity immediately after the impulse.

If ω_f is the final angular velocity, then the conservation of angular momentum requires that

$$L_f \;=\; L_i$$

$$I\omega_f \;=\; I_t\omega_i$$

$$\Rightarrow \qquad \omega_f \;=\; \frac{2(r+h)K}{r[(r+2h)m + (r+h)m']}$$

Notice that the final angular velocity is always less than that right after the impulse is delivered unless $m' = 0$ in which case the tank is empty.

The mechanical energy lost through friction is related to the temperature of the water by a quantity known as the specific heat. The ***specific heat*** of a material is defined as that amount of energy needed to raise the temperature of a unit volume of the material by one degree. These numbers are tabulated for different materials and are fairly constant over normal temperature ranges.

The actual calculations will not be carried out here because they are beside the point of this presentation. However, the following remarks are an outline of how it is done. First the lost energy ΔH is found by subtracting the final energy from the initial energy:

$$\Delta H \;=\; \frac{1}{2}I\omega_f^2 \;-\; \frac{1}{2}I_t\omega_i^2$$

Then the change in temperature ΔT of the water is found from

$$\Delta T \;=\; \frac{\Delta H}{V \cdot E}$$

where V is the volume of water, and E is the specific heat of water.

Note: Water is used as the standard for specific heats. Its specific heat is unity but the ΔH in the first of the two above equations must be converted to different units before it can be used in the second.

5.3 COMBINED TRANSLATIONAL AND CIRCULAR MOTION

5.4.1 The Decomposition of Combined Motions

The general motion of a body can become very complex. However, it can be analyzed as the vector sum of two simpler motions: the motion of the *CM* and the motion around the *CM*. All the concepts necessary to enable the analysis of this decomposition have now been gathered: treat the *CM* as a point mass *M* moving according to $\vec{F} = m\vec{a}$ where $\vec{F}$ is the net external force acting on the system; then analyze the motion of the system around the *CM* using the methods of the previous chapter.

More generally, it is not necessary to choose the *CM* as the representative point of the body. Any point will do. But the fact is that you will have to describe the motion in relation to an inertial laboratory system of coördinates and the process of switching from one coördinate system to the other is greatly simplified if the *CM* system is one of them.

Recall that the momentum in the *CM* system is always zero. That means that the momentum in the laboratory system is simply the momentum of the *CM*. Also the kinetic energy of motion in the laboratory system is the kinetic energy *in* the *CM* system plus the kinetic energy *of* the *CM* system. These relations are not true for any other point. (See *sections* **3. 2. 7. 2** *and* **3. 2. 7. 3**.) Thus it follows that if the motion of a body is decomposed into linear and rotational motions using any point other than the **CM** the process of switching from the laboratory coördinates to those centered at the chosen point becomes encumbered with more complex relations.

In the following section, we will analyze the following problem: find the motion of a rod which is initially at rest on a frictionless surface and is given an impulse **K** in a direction normal to its length. The problem itself is simple and straightforward but it a doorway to some interesting questions.

First, it will be seen that there is one point along its length which is stationary for the first instant after the impulse (we are talking about an impulse delivered in a very short time interval). This point is called the **instantaneous center of rotation** and is reminiscent of the original program envisioned whereby the general motion of a system could be decomposed into components tangential and normal to the path of motion. Recall that this program was abandoned because finding the center of rotation to the path of motion was too complex to lead to easy solutions. The problem in question will illustrate this fact.

Nevertheless, this decomposition which is natural to Newton's law of dynamics (recall that the dynamics law separates the component which alters the path from that which alters the speed) still offers insights into the nature of the motion of mechanical systems. In particular, it will be seen that the instantaneous center of rotation is independent of the strength of the impulse: it depends only on the geometry of the body and the point at which it is struck.

"So what?!" you say. Well, think about it for a moment. The implication is that the two points, the one where the impulse was delivered and the other which does not move for the first instant are in some way characteristic of the body itself: the points are related in a sense which we will analyze. They are called *conjugate points* of the body; their relation to each other is of considerable interest. For example, if the impulse is delivered at a point *A* and a point *B* is the instantaneous center of rotation, then *A* would be the instantaneous center of rotation for an impulse delivered at *B*. In other words the two conjugates interchange roles.

Another consideration which leads us to suspect such an analysis might be fruitful is as follows. The deliverance of an impulse is among the simplest methods to trigger the combined translational and rotational motion under inspection. In the ensuing motion which is free from outside influences, the translational motion is the reaction to object's resistance to a change in *linear* motion while the rotational motion is the reaction to the object's resistance to a change in *circular* motion. These resistances are respectively the mass and the moment of inertia of the body. The expectation is that the instantaneous center of rotation which in a sense compares the two component motions will reveal some of the relations between the mass and the moment of inertia. These quantities are a fun.ction of the body and its geometry.

Baseball bats, golf clubs, hammers, etc. are designed to be gripped at the point conjugate to the point where the impulse will be delivered because this is the point that does not react to the impulse. Hence if the implement is gripped in the right place, there is no 'sting' felt in the hand. (If one were to hammer all day holding the hammer too close to or too far away from the peen, the result would be a blistered hand.) Also note that since the position of the conjugate point is independent of the strength of the impulse, the 'sting' is minimized at the proper grip position regardless of how easy or hard the hammer-strike.

A second interesting property of conjugate points is that if a body is suspended from one of them and allowed to swing like a pendulum, the period of the oscillation will be the same as that which would result if the body were suspended from the other one.

This particular phenomenon can be carried further in that it relates to what is called a simple pendulum. A *simple pendulum* is one composed of a point mass suspended from a massless string. Its period of oscillation depends only on l/g, the ratio of its length to the acceleration due to gravity. In addition to the fact that a physical body exhibits the same period when suspended from either of two conjugate points, the length of the simple pendulum having that same period is equal to the distance between the conjugates.

Although the simple motion of the rod is used as an indicator of these phenomena, they are general properties of rigid bodies. Once motivated by the rod analysis, these features will be examined in a more general context.

Enough suspicions about, and advertisements of things to come. It's time to see if all this really happens.

5.4.2   Conjugate Points of a Rigid Body

5.4.2.1   The Motion of a Thin Rod under the Action of an Impulse

A thin rod of length l and mass m is at rest on a frictionless surface. An impulse K is delivered laterally to the rod at a distance x from its CM as shown in $figure\ 5.6.a$. What is the ensuing motion of the rod?

In the following, v is the speed of the CM, ω is the angular velocity around the CM and I_o is the moment of inertia around the CM.

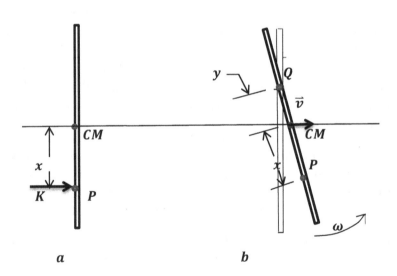

figure 5.6

Conjugate Points of a Rigid Body

For the motion of the **CM:**

$$Impulse \; = \; Change \; of$$
$$Momentum$$

$$K \; = \; mv$$

$$\Rightarrow \quad v \; = \; K/m$$

Hence, the **CM** of the rod moves to the right at a speed v.

For the motion around the **CM**:

$$Angular \; Impulse \; = \; Change \; in$$
$$Angular \; Momentum$$

$$Kx \; = \; I_o\omega$$

$$\omega \; = \; \frac{Kx}{I_o}$$

Hence, the angular velocity around the **CM** is ω in a counter-clockwise direction.

This completely describes the motion of the rod after the impulse. **Figure 5.6.b** illustrates how the translational and rotational motions add at the first instant after the impulse is delivered. From the diagram, it is clear that the geometry of the combined motions singles out a point **Q** which does not move initially. This point is the instantaneous center of rotation, so-called because the initial motion of the rod is a pure rotation around it. The rod however will not continue to rotate around **Q**: at the next instant it will exhibit pure rotation around a different point in the vicinity of **Q**. (The general difficulty of tracking the instantaneous center of rotation is the primary factor that limited the use of the tangential-normal decomposition as a general approach to mechanics problems.)

The task now is to find the distance y from the CM to Q. The condition to be satisfied is that the angular velocity ω times its moment arm y exactly cancels the velocity v of the CM:

$$\omega y = -v$$

Collect the last three equations:

$$K = mv \qquad (1)$$

$$Kx = I_0\omega \qquad (2)$$

$$\omega y = -v \qquad (3)$$

It is now possible to obtain a relation which does not depend on the original impulse K: substitute (1) into (2);

$$mvx = I_0\omega. \qquad (2')$$

By using (3), we obtain

$$-m\omega xy = I_0\omega \qquad (2'')$$

$$\Rightarrow \quad xy = -\frac{I_0}{m}$$

This last equation shows that the product of x and y is a constant. The equation has 4 interesting properties:

1. *it is independent of the impulse K and the body's reaction to it. this implies it is a function only of the body;*
2. *it is independent of the mass of the body (since the moment of inertia is always proportional to the mass). this implies that it depends only on the mass distribution and the geometry of the body. when the mass distribution is uniform throughout the body, it depends only on the geometry;*

3. *it is symmetric in x and y which implies that their roles can be interchanged;*
4. *the product of x and y is negative which indicates that the points P and Q are always on opposite sides of the CM.*

5.4.2.2   The Definition of Conjugate Points

We note first that the above analysis applies to any rigid body. The diagram of *figure* **5.7** shows an arbitrary body subject to an impulse K. the analysis is precisely the same as that carried out above as long as the points P and Q are taken along that line through the CM which is perpendicular to the impulse.

Note that the points P and Q may be either inside or outside the body. The problem of delivering an impulse to a rigid body at a point outside its boundaries is not an analytic one; it is a problem in implementation. For example, the body may have attached to it a light (i.e. massless) rod which extends to the point where the impulse is to be delivered.

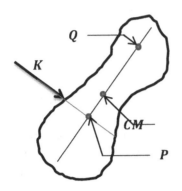

figure 5.7

Definition of Conjugate Points

We make the following ...

> *Definition:* **two points P and Q are conjugate points of a rigid body if the line joining them passes through the CM of the body and the product of their distances from the CM is I_0/M where I_0 is the moment of inertia of the body around its CM and M is its mass.**

It would be expected that the mechanics of a rigid body exhibit some symmetries and anti-symmetries with respect to this conjugation and this is in fact the case. The following symmetry will now be investigated: the period of oscillation exhibited by a rigid body when it is suspended from a point P and allowed to swing as a pendulum is the same as that exhibited by the body when it is suspended from the point Q conjugate to P.

This analysis will be facilitated if we find ...

5.4.2.3 The Period of a Simple Pendulum

A simple pendulum is one consisting of a point mass suspended from a massless string. The following analysis will show that the period of a simple pendulum depends only on the length of the string and g, the acceleration due to (local) gravity for the particular planet you are on.

Figure 5.8 shows a simple pendulum with point mass m and string length l. The analysis is restricted to small angular displacements because it is possible to find an explicit expression for the period in this case. However, the analysis applies more generally as will be seen in the next section.

The torque equation for motion around P is

$$\sum \tau \;=\; I\alpha$$

$$mgl\, sin(\theta) \;=\; ml^2 \frac{d^2\theta}{dt^2}$$

$$\Rightarrow \qquad \frac{d^2\theta}{dt^2} - \frac{g}{l}\theta \;=\; 0$$

(The small angle approximation $sin(\theta) \approx \theta$ has been used.)

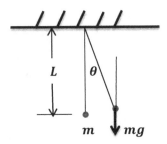

figure 5.8

Simple Pendulum

This is the same differential equation as the one obtained for simple harmonic motion. Its solution is

$$\theta = A\sin\left(\sqrt{\frac{g}{l}}t\right) + B\cos\left(\sqrt{\frac{g}{l}}t\right)$$

(See *section 3.3.3.3.*) The results obtained there indicate that

$$T = \frac{1}{2\pi}\sqrt{\frac{l}{g}}$$

The period of a simple pendulum is thereby related to the single parameter *l*, the length of the string.

In the following section, the period for a physical pendulum rotating around a point *P* will be determined and it will be found that the having the same period has a string length equal to the distance from *P* to its conjugate point *Q*. It will then follow that if the body is suspended from point *Q*, it will oscillate with the same period.

5.4.2.4 The Period of a Physical Pendulum

Figure 5.9 is a diagram of a physical pendulum suspended from a point **P** which is displaced from the **CM** by a distance **x**. The point **Q**, conjugate to **P** is shown at a distance **y** from the **CM**. From previous results, it is known that $xy = I_0/M$, where I_o is the moment of inertia around the **CM** and **M** is the mass of the body. It will now be shown that the length of the simple pendulum having the same period is $(x + y)$, the distance between **P** and **Q**.

The torque equation around **P** is

$$\sum \tau = I\alpha$$

$$Mgx\,sin(\theta) = I_p \frac{d^2\theta}{dt^2}$$

Rearranging terms and using the small angle approximation as before.

$$\frac{d^2\theta}{dt^2} - \frac{Mgx}{I_P}\theta = 0$$

The period of the oscillation is

$$T = \frac{1}{2\pi}\sqrt{\frac{I_P}{Mgx}}$$

By comparison to the period for a simple pendulum, it is seen that the string length l of the simple pendulum having the same period is

$$l = \frac{I_P}{Mx}$$

Use the parallel axis theorem to get this length in terms of I_0, the moment of inertia around the **CM** ($I_P = I_0 + Mx^2$):

$$l \;=\; \frac{I_0}{Mx} + x$$

Use the relation

$$xy = \frac{I_0}{M}.$$

Then $\qquad l \;=\; y + x .$

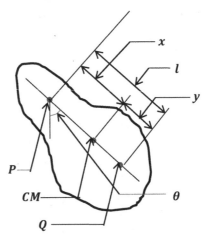

figure 5.9

Physical Pendulum

Hence, the length of the simple pendulum having the same period is the distance between the point P and its conjugate Q. If the physical pendulum is now suspended from Q, the simple pendulum with the same period still has the same length. It follows that the period of the physical

362

pendulum when it is suspended from P is the same as that when it is suspended from Q.

It was not necessary to use the small angle approximation to prove this. Consider the **unsolved** differential equations for both the simple pendulum and the physical pendulum:

for the simple pendulum:

$$\frac{d^2\theta}{dt^2} - \frac{g}{l}sin(\theta) = 0;$$

for the physical pendulum:

$$\frac{d^2\theta}{dt^2} - \frac{Mgx}{I_P}sin(\theta) = 0.$$

These are the equations of motion without the small angle approximation.

If we set

$$l = \frac{I_P}{Mx}$$

these two differential equations are identical and, for the same initial conditions, have identical solutions. Therefore, the periods of oscillation for the two different pendulums are the same under the same initial conditiions (even if we don't have an explicit expression for them).

5.4.2.5 The Natural Units of Length of a Rigid Body

Attention is now restricted to two dimensional plates. The axis through the *CM* will be the one normal to the plate. Consider the relation

$$xy = \frac{I_o}{M}$$

relating the distances of two conjugate points from the **CM**. The right side of the equation is constant. Therefore, each point on a circle of radius x centered at the **CM** is conjugate to a point on a circle of radius y centered at the **CM**. (These two circles are the collection of all points from which the body can be suspended to produce the same period of oscillation.) The two circles might be thought of as conjugate to each other. There is one particular circle however, that is its own conjugate: this is the case when $x = y$. The radius λ of the self-conjugate circle is

$$\lambda = x = y = \sqrt{\frac{I_o}{M}}$$

The length λ is a natural unit of length for the body in question. Notice that

$$I_o = M\lambda^2 ,$$

which indicates that a point mass M at a distance λ from the axis has the same moment of inertia as that of the given extended mass.

Also note that if x and y are expressed in terms of the natural unit λ

$$\zeta = \frac{x}{\lambda} \quad \text{and} \quad \eta = \frac{y}{\lambda},$$

then the equation

$$xy = \frac{I_o}{M}$$

in terms of the new unit is

$$\left(\frac{x}{\lambda}\right)\left(\frac{y}{\lambda}\right)\lambda^2 = \frac{I_o}{M}$$

$$\zeta \eta \, \lambda^2 \;=\; \lambda^2$$
$$\Rightarrow \quad \zeta \eta \;=\; 1 .$$

But ζ and η are the distances expressed in terms of the characteristic length λ of the body from the *CM* to each of a pair of conjugate points. Using λ as the unit of measurement, the distances from each of the two conjugate points *P* and *Q* to the *CM* are simple reciprocals of each other.

It is interesting to note in passing that in three dimensions, the moment of inertia of a body is represented by a 3×3 tensor. From this tensor, three directions together with a particular number for each can be determined. These directions and numbers are characteristic of the inertial effects of the body. If the directions are used as the principal directions for a coördinate system and the characteristic numbers, call them λ_x, λ_y and λ_z, are used to scale the variables in each of these directions (as was done above), then the result is to clear the algebra of many constants which would otherwise have to be carried along throughout the analysis. The three λ's are similar to the λ above which was used to scale the *x*- and *y*-directions: they are a set of three natural units each associated with one of the three directions.[20]

This approach is most useful when an extensive and detailed study of the geometry and motion of *one particular* rigid body is to be carried out. An example

---

[20] Each moment of inertia matrix has associated with it three characteristic vectors each linked to a characteristic number. They are known as eigen-vectors and eigen-values. Their study is invaluable in that it provides a theory for the determination of parameters which decompose complex multi-dimensional problems into simpler one-dimensional component problems. This study is referred to as 'spectral theory' and is a particular topic encountered in matrix algebra. The current analyses are the restricted version of this topic which applies to the restricted motions considered here.

would be the design, manufacture and evaluation of a particular gyroscope.

There will be no call to set up any such set of units here.

5.4.3 Historical Note

On November 5, 1987, a cave with markings from pre-history was discovered in central Yugoslavia by an expedition of the National Geological Society of Great Britain. To the surprise of Sir George Wellington, who is a senior member of the board of directors and who has studied the markings for well over two decades now, the markings were a legacy left to us by a man who called himself Gaargh and who as luck would have it, was the designer of the first hammer. The following transcription of the markings is reprinted here, compliments of the *NGSGB*.

"My recent discovery, the bhram (hammer) has been a source of great consternation to me. While it is a most useful tool, its use leaves my hands so blistered and heat-sensitive that it is impossible for me to enjoy my favourite dish (instep of brontosaurus en brochette), as I have to hold the hot morsels in my hands to eat them. I am very motivated therefore, to redesign the bhram to correct the difficulty.

"The new design takes advantage of a theory recently advanced by my friend and co-worker Jareegha: a theory which is primarily concerned with the position of the gripping point along the garhepa (handle) of the bhram. According to the theory, the sting felt at the point of gripping can be minimized. Etched below is a diagram of the bhram and the calculations for the new garhepa design.

The issue is to place the gripping point at a position along the garhepa so that it satisfies the equation

$$xy = \frac{I_o}{m'}$$

where I_o is the moment of inertia of the bhram around its mass center, m' is the total mass of the bhram, x is the distance from the doka (peen) at P to the mass center and y is the distance from the mass center to the gripping point at Q.

"The mass M of the doka is large and concentrated in a small region at one end of the garhepa. The mass m of the garhepa on the other hand is distributed more or less uniformly along its length L.

"The distance x is determined by requiring that the mass moments around the CM add to zero (see *figure 5.10*):

$$Mx - m\left(\frac{L}{2} - x\right) = 0$$

$$\Rightarrow \quad x = \frac{L}{2}\left(\frac{\mu}{1+\mu}\right)$$

where $\mu = m/M$.

The distance x' is:

$$x' = \left(\frac{L}{2} - x\right)$$

which after simplification is

$$x' = \frac{L}{2}\left(\frac{1}{1+\mu}\right)$$

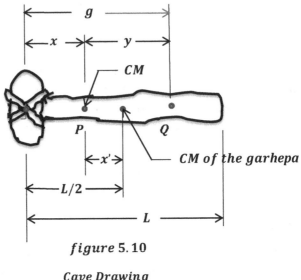

figure 5.10

Cave Drawing

"The moment of inertia around the mass center is found by adding the separate contributions of the doka and the garhepa. The contribution from the doka is found by use of the parallel axis theorem (to be discovered by Lagrange in 1783 AD):

$$I_d \;=\; I_{d_0} \;+\; Mx^2$$

$$I_d \;=\; I_{d_0} \;+\; \frac{ML^2}{4}\left(\frac{\mu^2}{(1+\mu)^2}\right)$$

where I_{d_0} is the moment of inertia of the doka around its own mass center. (I always use the same size doka so I already know the value of I_{d_0})

"The moment of inertia of the garhepa is found using the same theorem:

$$I_g \;=\; I_{g_0} \;+\; mx'^2$$

$$I_g = \frac{mL^2}{12} + \frac{mL^2}{4}\left(\frac{1}{(1+\mu)^2}\right)$$

"Add the two contributions together to get the moment of inertia I_o of the bhram around its **CM**. After some simplifying, it is

$$I_o = I_{d_0} + \frac{ML^2}{4}\left(\frac{1+\mu^2}{(1+\mu)^2} + \frac{\mu}{3}\right)$$

"I can now use Jareegha's theory to find the distance **y** from the **CM** to the gripping point:

$$xy = \frac{I_o}{(M+m)} = \frac{I_o}{M(1+\mu)}$$

$$xy = \frac{I_{d_0}}{M(1+\mu)} + \frac{L^2}{4}\left(\frac{1+\mu^2}{(1+\mu)^3} + \frac{\mu}{3(1+\mu)}\right)$$

Divide the last equation by

$$x = \frac{L}{2}\frac{\mu}{(1+\mu)}:$$

$$y = \frac{2I_{d_0}}{ML\mu} + \frac{L}{2}\left(\frac{1+\mu^2}{\mu(1+\mu)^2} + \frac{1}{3}\right)$$

$$y = \frac{2I_{d_0}}{ML\mu} + \frac{L}{2}\left(\frac{1+\mu^2}{\mu(1+\mu)^2}\right) + \frac{L}{6}$$

Add the above equation to **x**:

$$x + y = \frac{2I_{d_0}}{ML\mu} + \frac{L}{6} + \frac{L}{2}\left(\frac{\mu}{1+\mu} + \frac{1+\mu^2}{\mu(1+\mu)^2}\right)$$

"(Making cave etchings is a long and tedious process. Therefore the missing algebra steps are left as an exercise for those who read these walls.)

"Finally, the distance **g** from the doka to the gripping point is:

$$g \;=\; x + y$$

$$g \;=\; \frac{2I_{d_0}}{ML\mu} + \frac{L}{6} + \frac{L}{2}\left(\frac{1 + 2\mu^2 + \mu^3}{\mu(1+\mu)^2}\right)$$

"The new design using Jareegha's theory is greatly superior to the previous one. If you order within the next moon-cycle, you will receive two (2) bhrams for the barter of one.

"May the farce be with you.

Gaargh,

Inventor extraordinaire."

Gaargh, who was very much ahead of his time, unknowingly used the theory of conjugate points. His analysis is correct and the new design fixes the gripping point of the hammer (bhram) at the point conjugate to the peen (doka).

In appreciation of this valued legacy, we wish Gaargh and Jareegha 'bon appetit'.

Incidentally, in the context of the hammer or any other rigid body to which a sharp blow is to be delivered, the gripping point is now known as the ***center of percussion.***

5.4.4 Sample Problems in Combined Translational and Rotational Motions

Sample Problem 1:   **(Combined Motion of Two Attached Disks)**

Two identical disks of radius r and mass m are free to slide on a frictionless table top. They collide inelastically so that their centers are a distance a ($a > 2r$) apart after collision. This is accomplished via a massless 'catcher' attached to the mass initially at rest as shown in the diagram of *figure* **5.11.** *a.* The other disk has an initial velocity $\vec{v}$ so that the *CM* of the system moves along the negative **x**-axis toward the origin. Describe the motion after collision. (This problem was postponed from **section 3.2.2.**)

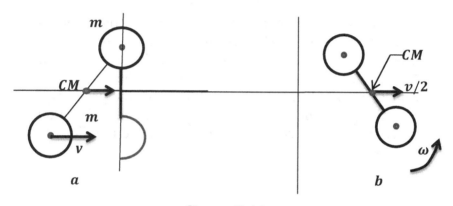

figure **5.11**

Combined Translational and Rotational Motion

Analysis

There are no external forces or torques on the system. Therefore, both linear and angular momenta are conserved. The latter can be computed around any point but the simplest and clearest analysis results from the decomposition into the motion of the *CM* plus motion around the *CM*. Except for finding the moment of inertia of the final configuration around the *CM*, this problem is solvable by inspection.

(Note that if the masses collide perimeter to perimeter instead of via a 'catcher' , the problem is essentially the same. The only difference is the geometry problem to determine the angle of the line of centers at the time of collision.)

For the linear motion, we have by inspection that the **CM** is initially moving with velocity $v/2$ along the x-axis in the positive direction. This will be the same after collision since momentum is conserved.

For the motion around the **CM**, the initial angular momentum L_o in the **CM** system can be found. It is the same after collision since angular momentum is conserved. The angular velocity ω after collision is then computed from $L = I\omega$. All that is required to solve the problem is to find I_o the moment of inertia around the **CM** after collision. This is accomplished by use of the parallel axis theorem.

Solution

In the **CM** system, one mass moves with velocity $v/2$, the other with a velocity **of** $-v/2$, both in the x-direction. The angular momentum is found from

$$\vec{L} \;=\; \vec{r} \times \vec{p}.$$

The position vectors are from the **CM** to each of the bodies. When crossed into each of the linear momenta, it is seen that both cross products point out of the page. The total angular momentum then is

$$|\vec{L}| \;=\; \frac{mva}{2}$$

Let I_o be the moment of inertia of one disk around its own center. Then by the parallel axis theorem (for one disk)

$$I' = I_0 + MR^2$$

$$I' = \frac{1}{2}mr^2 + m\left(\frac{a}{2}\right)^2$$

The total moment of inertia around the *CM* after collision is twice that amount:

$$I = 2\left[\frac{1}{2}mr^2 + m\left(\frac{a}{2}\right)^2\right].$$

Because angular momentum is conserved,

$$L_f = L_i$$

$$I\omega = mv\frac{a}{2}$$

$$\frac{m}{2}(2r^2 + a^2)\omega = mv\frac{a}{2}$$

$$\Rightarrow \qquad \omega = \frac{va}{2r^2 + a^2}$$

Hence, after collision, the *CM* moves along the *x*-axis to the right with velocity *v*/2 and the disks rotate around the *CM* with constant angular velocity ω given by the last expression. This completely characterizes the motion. (See *figure* 5.11.b.)

Sample Problem 2

A solid sphere of radius *r* and mass *m*, moving under the local gravity force, slides (or rolls off) a fixed hemispherical shell as shown in *figure* 5.12. If it starts at rest from the top, determine the angle with the vertical at which the sphere leaves the surface of the hemispherical shell both when the contact is frictionless and when the sphere rolls without slipping.

The only external force on the system is the local gravity force. Therefore energy is conserved. Hence, the speed and in the case of rolling, the angular velocity of the sphere can be determined as a function of the height at any point during the motion. The condition that the sphere stays on the surface of the hemisphere is that it exhibit circular motion. For this it requires a radial component of force (centripetal). This force is initially provided by the radial component of the weight: any excess thus provided is neutralized by the normal force.

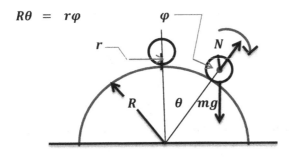

figure 5.12

A Sphere Rolling Off a Hemisphere

As the sphere moves down the surface of the hemisphere however, its speed increases requiring more centripetal force to maintain the circular motion and simultaneously, the weight has a smaller and smaller component in the radial direction. At some point these factors gang up and there is insufficient radial component of force available to the sphere for it to maintain circular motion and it leaves the surface of the hemisphere. The condition for this event is that the normal force equal zero. (Recognizing this fact is perhaps the most difficult part of the problem.) At this point of the motion, there is no more excess radial component of the weight available and in the next instant the increased speed of the sphere will cause it to deviate from circular motion.

In the case of rolling, the constraint condition is that the rotation of the sphere be such that the arc length rolled out is equal to the arc length traversed on the hemisphere. This constraint relates the velocity of the *CM* of the sphere to its angular velocity around the *CM*.

Solution

The rolling situation will be analyzed first and as we go along, an artifice will be introduced which will produce the solution for both the rolling and sliding (frictionless) cases.

First, obtain an expression for the kinetic energy. Let the subscript '*t*', refer to 'translational' motion, and the subscript '*r*', to rotational motion. The total kinetic energy *T* is the kinetic energy of the *CM* plus the kinetic energy around the *CM:*

$$T = T_t + T_r$$

$$T = \frac{1}{2}mv^2 + \frac{1}{2}I\omega^2$$

where v is the speed of the *CM*, *I* is the moment of inertia around the *CM* and ω is the angular velocity of the sphere around its *CM*. The velocity and the angular velocity are not independent however: they are related by the constraint that the surfaces stay in contact without slipping. The equation of constraint is

$$R d\theta = r d\varphi,$$

where θ is the angular displacement along the hemisphere and φ is the angle through which the sphere rotated to get there. Dividing this last relation by dt, it follows that

$$\frac{d\theta}{dt} = \frac{r}{R}\frac{d\varphi}{dt}$$

The velocity v of the sphere is related to the angle θ by

$$v = (R+r)\frac{d\theta}{dt}$$

and hence, to the angle φ by

$$v = \frac{r(R+r)}{R}\frac{d\varphi}{dt}.$$

But $d\varphi/dt$ is the angular velocity ω of the sphere around its center. Therefore, ω and v are related by

$$\omega = \frac{Rv}{r(R+r)}$$

This is the case when the sphere rolls. When it slides, $\omega = 0$. The artifice mentioned above is introduced at this point: the above expression is multiplied by a quantity q with the understanding that later q will be set equal to unity when the sphere is rolling, and it will be set equal to zero when the sphere is sliding. Thus both answers will be produced from the single final equation.

Using the above result and introducing the quantity q, the expression for the kinetic energy is

$$T = T_t + T_r$$

$$T = \frac{1}{2}mv^2 + \frac{1}{2}I\omega^2$$

$$T = \frac{1}{2}mv^2 + \frac{1}{2}\left(\frac{2}{5}m r^2\right)\left(\frac{qR^2v^2}{r^2(R+r)^2}\right)$$

After simplifying, $T = \left(\dfrac{1}{2} + \dfrac{qR^2}{5(R+r)^2}\right)mv^2$

The equation for the conservation of energy taking the height equal to zero at the center of the hemisphere is

$$H_i = H_f$$

$$mg(R+r) = mg(R+r)\cos(\theta) + T$$

$$mg\,(R+r)\,[1 - \cos(\theta)] = \left(\dfrac{1}{2} + \dfrac{qR^2}{5(R+r)^2}\right)mv^2$$

$$mg\,[1 - \cos(\theta)] = \left(\dfrac{1}{2} + \dfrac{qR^2}{5(R+r)^2}\right)\dfrac{mv^2}{(R+r)}$$

The last factor on the right (if you recall) is the centripetal force. It is the force necessary to maintain the circular motion of the sphere and is provided by the radial component of the **mg** force.[21] Hence it is a condition of the problem that this factor equal the radial component of force while there is contact.

From the diagram, it is seen that the radial component of force pointing toward the center of the hemisphere is

$$F_r = mg\cos(\theta) - N.$$

Substitute this expression for the rightmost factor of the previous equation:

$$mg\,[1 - \cos(\theta)] = \left(\dfrac{1}{2} + \dfrac{qR^2}{5(R+r)^2}\right)[mg\cos(\theta) - N].$$

This equation is true for all angles as long as the surfaces are in contact. The condition for the angle at which the sphere leaves the surface of the hemisphere is $N = 0$. It follows that the equation for the critical angle is

---

[21] If you don't recall, you'll still get the answer, but it's more work that way.

$$1 - cos(\theta) = \left(\frac{1}{2} + \frac{qR^2}{5(R+r)^2}\right)cos(\theta)$$

Solving for $cos(\theta)$, the angle at which the surfaces lose contact must satisfy the equation

$$cos(\theta) = \frac{10(1+\frac{r}{R})^2}{15(1+\frac{r}{R})^2 + 2q}$$

By setting $q = 1$, this becomes the result when the sphere rolls without slipping; by setting $q = 0$, this becomes

$$cos(\theta) = \frac{2}{3}$$

which is the result when the surfaces are frictionless.

Exercise

Parallel the solution of the previous problem when the objects are cylinders of radii of r and R instead of spheres. (Assume the axes of the cylinders are parallel throughout.) You should obtain the solution.

$$cos(\theta) = \frac{4\left(1+\frac{r}{R}\right)^2}{6\left(1+\frac{r}{R}\right)^2 + q}$$

When $q = 1$, this is the solution for rolling.

Discuss why the frictionless case ($q = 0$) still produces the answer

$$cos(\theta) = \frac{2}{3}$$

5.5   D'ALEMBERT'S PRINCIPLE

5.5.1   The View from an Accelerated Coördinate System

D'Alembert's principle addresses the issue of viewing dynamics from an accelerated coördinate system. The rules that apply are simple but they tend to get confusing unless you are clear from the outset about adopting an accelerated coördinate system as your home and sticking to it. When you say you are in some particular coördinate system, you mean that you are at rest with respect to it. If the coördinate system happens to be inertial, there is no problem: all the laws of dynamics apply as previously described. If however, your coördinate system is accelerating, this is no longer the case.

As an example, imagine yourself on a train as it pulls away from the station with some acceleration $\boldsymbol{a}$. (Note: moving with constant velocity doesn't make it for these considerations: we are talking about **accelerated** coördinate systems.) Your friend outside describes your situation as acceleratory and says appropriately that your contact force with the train (either the seat or the floor or a strap you are hanging onto) exerts a net force on you causing you to accelerate and hence, stay with the train. This is the viewpoint from an inertial frame (at least as much as the earth can be considered as such). He says the force acting on you and your acceleration are related by good old

$$\vec{F} \;=\; m\vec{a}.$$

Because you are used to trains and accustomed to thinking of the earth as an inertial system, you can look out the window and adopt *exactly the same viewpoint*. In this case, you are also viewing your own situation from an inertial frame even though you don't happen to be in one. You have not adopted the train system as your home. If however, you consider the train (instead of the earth) as your system, you regard yourself as having zero acceleration because you are *at rest with respect to the train.* Your paradox is this: you have a net force acting on you, but nevertheless you are not accelerating: you are and remain at rest.

What happened to $\vec{F} = m\vec{a}$? It says zero acceleration means zero net force: but does it?! Obviously not. Your friend sees that very clearly from the station. Should the laws of physics be thrown out the window back into the inertial frame of the station? No such luck! The task here is not to avoid the discrepancy but rather to explain it. The fact is that the laws of dynamics are still in tact: but it is necessary that you 'fess up!': you have not met the 'in-an-inertial-frame-of-reference' part of your tacit agreement with Sir Isaac because your frame is *not* inertial. Consequently it is necessary that you account for the acceleration of your frame of reference.

If you adopt the train as your home then you must be consistent: you must describe all motions relative to you and the train, and you must admit that everything at rest (or not) in your system is acted upon by a mysterious force whose origin is unknown, whose magnitude is *ma* and whose direction is toward the back of the train. Other forces on objects in your system may be tractable and understood but this particular one is not understood: it is universally present and affects every object. Thus as the train pulls away from the station, each passenger sinks back into his seat, you have to hang on to the strap to keep this force from pulling you back and making you fall. This mysterious force seems to act just like a local gravity force[22] but it points in the wrong direction. Mathematically, you say

$$F - ma = 0$$

That is to say, your acceleration is zero (because the right side of your dynamics equation is zero) but in addition to all the other forces in your system (contact, electro-dynamic, gravitational or what-have-you which are included in the 'F' of your equation), there is this ubiquitous $(-ma)$ force. If you were to make measurements in this system it would eventually become clear that

> *your entire system of mechanics is based on the premise that this force acts on every mass in much the same way as the local gravity force, mg.*

The essence of D'Alembert's principle is contained in the following question and its answer. The question is, "Why does the change from $F = ma$ to $F - ma = 0$, which is so mathematically trivial have such a profound impact on the physics involved?"

The answer stems from the fact that each of these equations is more than just a mathematical equality. Classically, the equation $F = ma$ was taken to imply that the ***causal*** factors of a mechanical motion are to the left of the equal sign and the ***reactions of the system*** are to the right, i.e ***causes on the left; effects on the right.*** That is the underlying nature of a dynamics equation. It was understood that way when it was first introduced but this basic understanding usually fades with time and use.

---

[22] It is interesting to note that the theory of general relativity makes this similarity into an equality by stating that inertial mass and gravitational mass are the same. By implication, this statement also draws an equivalence between gravitational fields and acceleration fields.

D'Alembert's principle requires that it be refreshed. By transposing the **ma** from the right side of the dynamics equation to the left, we are saying that this term is no longer **reactive** but **causal:** before it is transposed, it is linked with the kinematics of your system's motion; after transposition, it is connected to the forces which (collectively) empower the motions in your system.

Suppose now that the term has already been transposed. If there is a zero on the right side of your dynamics equation then according to your viewpoint there is no motion but there is this mysterious force which, because there is no apparent source of a push or pull associated with it, has been dubbed 'fictitious'. Stated a little differenty, the transposition of the **ma** term has given rise to both effects:

1. *the existence of the fictitious force;*
2. *the necessity to analyze all motions relative to your (accelerated) frame of reference.*

They are a set and hence, must enter and leave your considerations together.

Do not confuse 'fictitious' with 'ineffectual' or 'not really there'. If you try to stand without holding on to a strap or otherwise bracing yourself in the train as it leaves the station, the *'fictitious'* force, the force *'that isn't there'* will knock you on your backside. It is 'very much there' in your system.

This can be carried further. Note that only objects at rest in your system were considered. It is certainly possible for some objects to have an acceleration with respect to the train. In fact the entire universe can be viewed and measured from your accelerated system. The very station you are leaving (from your viewpoint if you are truly consistent) is accelerating toward the back of the train because there is no force counteracting the mysterious (*ma*) force that acts on it (as it acts on everything from your viewpoint).

What is the law of dynamics in your frame of reference? We'll refer back to your friend's inertial system so we're on more familiar ground. Your friend says the real acceleration of a body of mass *m*, is the acceleration *a* of the train added vectorially to the acceleration *a'* with respect to the train. Thus when the object is acted upon by a real net force *F*, he describes it by the equation

$$\vec{F} \;=\; m(\vec{a} + \vec{a}')$$

because the real acceleration as seen from the inertial system can be decomposed into $(\vec{a} + \vec{a}')$.

But your friend is free to choose any number of ways to decompose the real acceleration, in fact, he may choose not to decompose it at all. The above decomposition was chosen in deference to you: this is the particular one which enables him to relate accelerations in his inertial system to accelerations as seen from your non-inertial system.

From your viewpoint however, the acceleration is $\vec{a}'$, but the body is acted upon not only by the real net force *F* but also by this mysterious $(-ma)$ force that is everywhere in your system. Therefore, you include this $(-ma)$ term as a force and your description is

$$\vec{F} \;-\; m\vec{a} \;=\; m\vec{a}'$$

This is the dynamics equation in your accelerated system. The term on the right indicates that motions are to be viewed relative to yourself; the second term on the left indicates that you must include the fictitious force that takes your own acceleration into account. Once the

$(-ma)$ term is included as a force in your system, your dynamics equation is essentially the same as that for an inertial system.

One cannot help but notice that the difference between the two viewpoints is mathematically trivial and indeed it is this fact that tends to hide the dramatic physical difference implied by the above description. If the difference has not yet accrued enough dramatic content in your mind, it is hoped that what follows in the next few sections will remedy this.

It has just been shown that if Newtonian mechanics is to be consistent from one coördinate system to another, we must adopt ...

> *D'Alembert's Principle: the dynamics in an accelerated coordinate system are identical to the dynamics in an inertial system if in addition to the real forces acting, every object is considered to be acted upon by a fictitious force equal and opposite to its mass times the acceleration of the system.*

5.5.2 Examples of Accelerated Coördinate Systems

Systems Exhibiting Constant Acceleration

Consider the following. A train is traveling with constant acceleration a in the x-direction and a mass m is suspended from the ceiling by a string and adjusted so that it is at rest with respect to the train.

This arrangement will be described from both an inertial frame and the non-inertial frame which is accelerating with the train. This problem is obvious and simple but it is not the problem that is important. It is the development of the ability to change viewpoints from inertial to accelerating systems.

From the inertial frame:

Your friend at the station watches the train accelerating and notes that there are two forces acting on the mass, local gravity and the tension in the string. He draws a force diagram as shown in *figure* **5.13.a**. His viewpoint is that the mass has acceleration a in the x-direction and must therefore have a net force acting on it equal to ma. He writes the following equations to describe the system:

$$\left(\sum F_y = 0:\right) \qquad \Rightarrow \quad T\cos(\theta) \;-\; mg \;=\; 0$$

$$\left(\sum F_x = ma_x:\right) \qquad \Rightarrow \quad T\sin(\theta) \;=\; ma_x$$

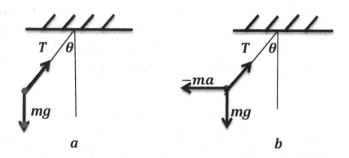

figure 5.13

DAlembert's Principle

His equations are the equations of dynamics: the net force acting is equal to the product of the mass and its acceleration a. *(2nd eq)*

From the accelerated frame:

You note that there are three forces acting on the mass, local gravity, the tension in the string and the mysterious $(-ma)$ force that acts on every mass in your system. *Simultaneously with this description, you describe the state of the mass as stationary.* Your viewpoint is that the mass is in static equilibrium and hence, the forces must add up to zero. Your force diagram is shown in *figure 5.13.b*; the equations you write to describe the system are

$$(\textstyle\sum F_y = 0): \qquad \Rightarrow \; T\cos(\theta) \; - \; ma \; = \; 0$$

$$(\textstyle\sum F_x = 0): \qquad \Rightarrow \; T\sin(\theta) \; - \; ma_x \; = \; 0$$

These are the equations of statics: the sum of all the forces acting on the mass is zero.

Obviously, both you and your friend upon solving your respective sets of equations will come up with the same results for T and θ. Since your equations and his differ so trivially in a mathematical sense, it is urged that you focus attention on the difference in viewpoints rather than the algebra.

Before proceeding to more complex accelerated frames of reference, it is emphasized that when you adopt an accelerated frame, you must do two things simultaneously:

1. *add the fictitious $(-ma)$ force to every object in your system;*
2. *refer all measurements to your system.*

If you were to go part way with this, i.e. do some but not all of the things necessary for consistency or in any way renege on the entirety of your commitment to view the universe from your accelerated frame, then there will be wrong answers, weeping and gnashing of teeth.

These ideas can be applied easily enough when the acceleration is constant. The trick is to hang on to this simplicity when the non-inertial frame exhibits a more general acceleration.

Spinning Your Keys on Their Chain

Consider another example. Suppose you are spinning your keys on the end of your keychain so that they revolve around your hand with constant speed in a horizontal plane. (The chain sweeps out a 'pointed hat' that's somewhat flattened, but this does not enter into the problem.) From an inertial frame, the description is familiar to us: the velocity vector is continually changing direction and hence, there must be a net force normal to the direction of motion. This (centripetal) force has been calculated before: it is directed radially inwards and has a magnitude given by

$$|\vec{F}| \;=\; \frac{mv^2}{r} \;=\; m\omega^2 r.$$

This force is provided by the radial component of the tension in the chain. The gravity force is negated by the vertical component of the tension. (See *figure* $5.14.a.$)

But what of the view from the accelerating system? ...i.e. from the system in which the keys are at rest. You have to go all the way now: imagine a tiny version of yourself sitting on the center of rotation (where your fingers are holding the chain); imagine further that you are spinning with the chain-key assembly so that the keys are always directly in front of you. You have secured your friend (also tiny) to the keys and equipped him with a movie camera so he can record what happens.

This is quite a different world you are in. The keys are at rest in front of you and the rest of the world beyond the keys is rotating (God knows why) at some angular velocity you measure to be ω, (the negative of the ω in the above equation which describes things from an inertial frame). You note that there is a tension in the chain tending to pull the keys toward you. You also note that gravity is pulling the keys downward. There is no other force acting which is obvious to you, so you can't quite understand why the keys are not accelerating towards you. They seem to be suspended in air, just sitting there while the landscape continues to gyrate around you and them.

You decide to investigate this phenomenon and proceed to climb out along the chain toward the keys. As you move away from your initial position, you notice that some mysterious force is pulling you farther out along the chain. The further you are from your initial position, the stronger the force seems to be. 'Oh, oh", you say, "I remember once when I was on a train. There was this mysterious force that kept trying to pull everything towards the back of the train. Now I'm in a system that's even worse; everything seems to be pulled away from the point where I started and the further I get from it, the stronger the pull. Man, this is weird!'

You figure that it's this mysterious force that's holding the keys out there suspended in space. So you start to measure this mysterious force. The first thing you notice is that there is a preferred point in your world, the point where you started. The force does not act there. Next you find that the force which acts on every mass in your world is proportional to both its mass and its distance from the preferred point. If **k** is the proportionality constant you measure, then you conclude that the force is of the form

$$|\vec{F}| \quad = \quad kmr\,.$$

What you have measured is the negative of the centripetal force. The **k** in your equation is the square of the angular velocity of your system. (However, you do not know that because you have adopted the belief that you are stationary. You would discover eventually that the **k** equaled the square of the angular velocity of the landscape as seen from your viewpoint.) This force is fictitious and is referred to as the ***centrifugal (center-fleeing) force.*** It is part of the fictitious *(-ma)* force acting on every mass in your accelerating frame.

Nevertheless you are unshakable in your pursuit of mechanical knowledge and you say, "this is a real drag having to climb out on the chain every time I want to make a measurement. I'm going to build a platform under the chain so I can move in two directions instead of just along the chain." You go out to a distance **r** from the preferred point holding onto the chain all the while, and you prepare to nail the first cross-strut together. But someone has slipped you an ill-designed hammer with a stinging reaction at the grip. You raise the hammer ... strike, and ... "Ouch! ..." you drop the hammer, let go of the chain and you are at the free-fall mercy of the mysterious force(s) in your system.

"Hey!" you say while you're in flight, "when I was hanging on, that force was pulling me toward the keys, but as soon as I let go and started moving in that direction, something else started pulling me to the side! I'm moving away from the point where I started and sideways away from the chain." All this in a very short time because soon you crash into the spinning landscape --- or was it spinning after you let go? You're not sure because whatever gyrations it was going through, stopped when you let go of the chain, or maybe you started gyrating with it? "Not sure!? What? I don't understand!"

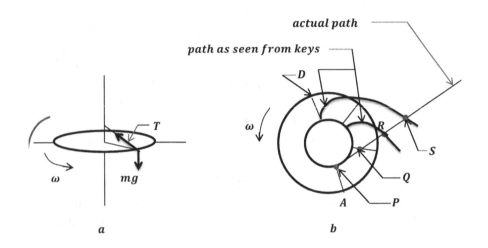

figure 5.14

Falling Off a Rotating System

A few crashes, a bounce or two and you have been returned somewhat cataclysmically to an inertial system. You note that you are now at rest with respect to the landscape and also that in spinning with it, there are no more mysterious forces. "Maybe the landscape is stationary and the chain was spinning", you conjecture. But, in the final analysis, you concede that there is no way to tell. All you really know is that the equations of physics are much tidier when there are no mysterious forces to deal with.

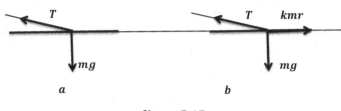

figure 5.15

Your Keys: a) from a Stationary System; b) from the Accelerated System

Your path is shown in the diagram of *figure* **5.14.b**. The curve indicated is as it appeared on the movie your friend took. It is your path as seen from the rotating system: say, you let go a *t = 0* when you were at point *P* and the keys were at *A*; later you were at *Q* and the keys were at *B*; later still, you were at *R* and the keys were at *C*. The motion picture taken by your friend traces out the curve shown from the circumference of the smaller circle to *R*; later the keys were at *D* and you were at *S*. The motion picture traces out a longer curve, from the circumference of the smaller circle to *S*. Then the landscape intervened.

Compare the following two descriptions of the dynamics: the first as seen from an inertial system and the second, as seen from the world into which you just ventured.

From the Inertial System

The keys are moving in a circle with angular velocity ω, and must therefore have a net force directed toward the center of rotation equal to

$$|\vec{F}| \;\;=\;\; m\omega^2 r.$$

There are two forces acting on the keys, the local gravity force and the tension in the chain. The vertical component of tension must be equal and opposite to the weight of the keys; the radial component must be equal to the above centripetal force which is necessary to sustain the circular motion. The force diagram is shown in *figure 5.15.a.* The accompanying equations are

$$\left(\sum F_y \;=\; 0\right): \qquad \Rightarrow \quad T\sin(\theta) \,-\, mg \;=\; 0$$

$$\left(\sum F_x = -m\omega^2 r\right): \;\Rightarrow \quad T\cos(\theta) \;=\; -m\omega^2 r$$

From the rotating system

The keys are not moving; they are in a state of static equilibrium. Therefore, the net force on them is zero. There are three forces acting on the keys: the local gravity force, the tension in the chain and the mysterious **kmr** force that pulls every mass in the system away from the preferred point. The vertical component of the tension must balance the weight of the keys; the radial component of the tension must balance the mysterious **kmr** force. The force diagram is shown in *figure 5.15.b.* the equations are

$$\left(\sum F_y \;=\; 0\right): \qquad \Rightarrow T\sin(\theta) \,-\, mg \;=\; 0$$

$$\left(\sum F_x + kmr = 0\right): \Rightarrow \; T\cos(\theta) \,+\, kmr \;=\; 0$$

Once again the two viewpoints result in equivalent equations and will produce the same results for T and θ.

It has been seen in the first example that if one adopts a system with constant acceleration, he can use all the laws of dynamics as long as he adds a fictitious (-ma) force to every mass. Conjoined to the inclusion of this force is the understanding that all measurements be referenced to the accelerating coördinates. In the second example, it was seen that the inclusion of a fictitious force called the centrifugal force enables the description of dynamics in a rotating system; as in the first case, the force is treated as if it acts on every mass in the system. Its magnitude F is

$$F \;=\; m\omega^2 r,$$

and it is directed radially outward. Once again, the inclusion of this force is accompanied by the understanding that all measurements be referenced to the accelerated system.

Recall that the Coriolis force acted only when there was a radial component of velocity in a rotating system: it is the side thrust necessary to increase your speed as you move farther away from the center of rotation. In the absence of this force, the movement will not be along a radial line because you cannot keep up with the angular movement (for a given non-zero angular velocity).

This is what happened when you let go of the chain: you started 'moving sideways away from the chain', you said. Viewed from an inertial system, you simply moved in a straight line as the angular movement of the system continued and 'got away from you'. (Gravity was pulling down, but this is a different direction and has no bearing on what we are talking about now. In *figure* **5.14.b**, the falling due to gravity is into the page.) From the viewpoint of the rotating system, there was yet another mysterious force acting: this one pulled you off the radial line to the side as soon as you acquired a radial component of velocity. This fictitious force is equal and opposite to the *Coriolis* force. It is 'the force without a name', but colloquially it is often referred to as the *Coriolis* force: you have to divine from the context whether the reference is to the real *Coriolis* force or its equal and opposite fictitious force.

The situation is similar to one encountered before. To see this, imagine that you are actually moving along a radial line. The view from an inertial frame is that there is a real force delivering a side thrust to adjust your speed to match that of your rotating system as your distance from the center changes. In the rotating system, you still experience this Coriolis force, but your explanation is different: as you move, you also experience the fictitious (equal and opposite) force. Hence, you say that movement along a radial line requires the Coriolis force to counteract the fictitious one acting in your system.

Each of the examples discussed demonstrated D'Alembert's principle: to view the universe from an accelerated coördinate system, the laws of mechanics remain in tact if all measurements are referred to the accelerated system and every mass is considered to be acted upon by a fictitious force equal and opposite to its mass times the acceleration of the system.

In the final analysis, the principle accounts for the inertial resistance of all masses to the acceleration of the system from which they are viewed. This is simple enough in itself. The difficulty is associated with the ability to adopt the viewpoint of being at rest in the accelerating system, and also to visualize motions in that system together with the same motions seen from an inertial system.

5.5.3 Sample Problems in Accelerated Systems

Sample Problem 1:   Equivalence of a Gravity Field and an Acceleration Field

A simple pendulum of length l and mass m is in a system that is moving parallel to the earth in the x-direction with constant acceleration a. What angle θ does the string make with the vertical when the mass is in equilibrium in the accelerated system? What is the tension T in the string when the system is in this position? What is the period S of small oscillations around this equilibrium?

Analysis

The equations can be written in either an inertial frame or the accelerated frame. In either case, it is a straightforward matter to find the tension and the angle. But there is a simpler approach to this problem. Consider the situation from the accelerated system and in particular look at the local gravity force and the fictitious $-ma$ force. The point was made earlier that gravitational fields and acceleration fields are equivalent (see footnote, *section* **5.5.1**). It is as if g and a are two components of a gravitational field of magnitude $\sqrt{a^2 + g^2}$ with a new 'downward' direction defined as shown in *figure* **5.16.b**. Since a and g are both constant, we have the equivalent of a gravity field of the above magnitude acting in the 'downward' direction, the one indicated in the figure. (It is apparent from the diagram that 'down' is defined as parallel to the vector sum of $\vec{g}$ and $\vec{a}$, and 'up' as anti-parallel to it.)

Having drawn this equivalence, the questions can be answered by inspection from the diagram: the equilibrium angle θ is determined by

$$\tan(\theta) \;=\; \frac{a}{g}\;;$$

the tension is determined by

$$T \;=\; m\sqrt{a^2 + g^2}\,,$$

(this is just the 'weight' in the equivalent 'local gravity' field); the period S of small oscillations around equilibrium is the analogue of

$$S \;=\; \frac{1}{2\pi}\sqrt{\frac{l}{g}}\,,$$

which is the period for a simple pendulum in the normal gravitational field.

Hence, it is

$$S \;=\; \frac{1}{2\pi}\sqrt{\frac{l}{\sqrt{a^2 + g^2}}}\,.$$

Incidentally, it is also possible to define a potential energy analogous to that of the local gravitational field:

$$V \;=\; m\sqrt{a^2 + g^2}\,z\,,$$

where **z** is distance in the **altered** up-down direction (**not** vertical).

This could have been seen by noting that the resultant of the gravity force and the fictitious $-\,ma$ force is a constant force field. (Refer to **sections 1.3.4.1** and **3.3.2.**) Although everything is lopsided by an angle θ, everything pertaining to constant force fields and in particular to the local gravity field, applies as long as you are faithful to the analogy drawn here.

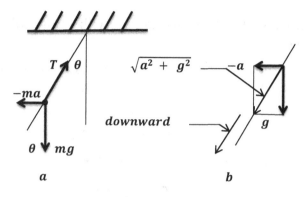

figure 5.16

The Equivalence between Gravity and Acceleration

We will now obtain the above results formally.

Solution

The equations of equilibrium come from the diagram of *figure* 5.13 and were already found in *section* 5.5.2. We have

$$T\cos(\theta) - mg = 0$$
$$T\sin(\theta) - ma = 0$$

Transpose the 2nd terms, square, add, and take the square root to obtain

$$T = m\sqrt{a^2 + g^2};$$

Divide the second equation by the first to obtain

$$tan(\theta) \;=\; \frac{a}{g}.$$

These last two results agree with those obtained by inspection.

To find the period of small oscillations around equilibrium, displace the mass by a small angle θ. The restoring torque is

$$\tau \;=\; m\sqrt{a^2 + g^2}\; l\, sin(\theta).$$

Substitute this in the torque equation:

$$\tau \;=\; I\alpha$$

$$m\sqrt{a^2 + g^2}\; l\, sin(\theta) \;=\; ml^2 \frac{d^2\theta}{dt^2}$$

Using the small angle approximation and rearranging terms, this equation becomes

$$\frac{d^2\theta}{dt^2} - \frac{\sqrt{a^2 + g^2}}{l}\,\theta \;=\; 0$$

This equation was solved in **section** $3.3.3.3$. Comparison with the equation there shows that the period of oscillation is

$$S \;=\; \frac{1}{2\pi}\sqrt{\frac{l}{\sqrt{a^2 + g^2}}}$$

This also agrees with the answer obtained by inspection.

Sample Problem 2: Constant Acceleration and Helium-Filled Balloons

You are in the same accelerated system as that of the previous problem and you are holding a helium filled balloon. Describe the equilibrium position of the balloon. Explain your answer.

Analysis and Solution

In the same way as the vector sum of $\vec{a}$ and $\vec{g}$ defines 'down' in the accelerated system, it also defines 'up' as the direction anti-parallel to it. Hence, the balloon hovers above your head leaning in the direction of the acceleration. The string makes an angle θ with the vertical (the same θ as in the previous problem).

This needs some explanation because our sense that everything is pulled backwards by the acceleration is offended by the fact that the balloon leans forward. This however, is really no different than the fact that it goes 'up' instead of 'down' in an inertial system. Note first that the equivalence of the system to a local gravity with 'up', 'down' and $\vec{g}$ redefined as above is trustworthy: the balloon rises in the normal gravity field and it also rises in this one, which is equivalent to it. And in fact the reason is the same: the helium is lighter than the air that it displaces.

In the accelerated system, the air that travels along with you is also subject to all the peculiarities that pervade your system. Hence, it is subject to a force directed backwards and because it outweighs the helium, it displaces the balloon forward. Our senses are offended by this result because the air is invisible to us; the balloon is not.

Discussion

The equivalence between the local gravitational field and the acceleration field that has been drawn here is an application of Einstein's hypothetical experiment of the man in the elevator (see *section* 3.3.4.1 for a description). The essential feature of the experiment is that from inside the elevator, it is impossible to distinguish between a gravitational field and an acceleration field.

The premise of general relativity is that this fact is extendable to all possible coördinate systems. Here we have extended it to coördinate systems having constant acceleration. The trustworthiness of the equivalence follows from a deep property of mass, namely, that inertial mass and gravitational mass are identical. Note that it was possible to draw the equivalence only because the mass in the (mg) force was the same as the mass in the $(-ma)$ force: the first of these is gravitational; the second is acceleratory (inertial)

.

Sample Problem 3: A Spring-Mass System on a Turntable

One end of a spring whose constant is k and equilibrium length l is fixed at a distance r from the axis of a rotating disk and a mass m is attached to the other end as shown in *figure* 5.17. Assume that the mass is constrained to move along a radial line and that it has frictionless contact. Find the equilibrium length of the spring as a function of the angular velocity of the disk. Show the spring exceeds its elastic limit and/or breaks when the angular velocity of the disk approaches the natural angular velocity of the spring-mass system.

400

Analysis

Viewed from the rotating system, this is a statics problem: the equilibrium length is determined by the condition that the (fictitious) centrifugal force be equal and opposite to the restoring force of the spring. (Refer to *figure 5.17*.)

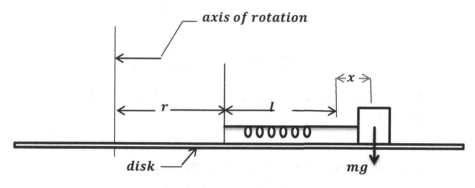

figure 5.17

Spring-Mass System on a Turntable

It will be seen that as the angular velocity of the disk approaches the natural angular velocity of the spring-mass system, the equilibrium length becomes infinite. In other words, the spring can no longer balance the centrifugal force at any position of the mass. This is a resonance effect which is not at all obvious.

Let x be the amount by which the spring is stretched. Then the equilibrium equation is

$$\sum F_x = 0$$

$$m\omega^2(r + l + x) - kx = 0$$

$$\Rightarrow \quad x \;=\; \frac{m\omega^2(r+l)}{k-m\omega^2}$$

Then the equilibrium length l' is

$$l' \;=\; l + x \;=\; \frac{m\omega^2 r + kl}{k - m\omega^2}$$

which becomes infinite when the denominator is zero:

$$k - m\omega^2 \;=\; 0$$

$$\Rightarrow \quad \omega \;=\; \sqrt{\frac{k}{m}}.$$

This is the natural angular velocity of the spring-mass system.

Sample Problem 4:   Weight at the Equator vs. Weight at the Poles

Assuming that the earth is spherical, show that any given mass weighs approximately 1/3 of a percent less at the equator than at either of the poles.

Analysis

The weight loss effect in question is a result of the fact that the earth is not an inertial system. A mass at a point on the equator is subject to a centrifugal force from the earth's rotation; at one of the poles, it is not.

Solution

Let w_p be the weight at one of the poles and w_e, the weight at the equator. Then

$$w_p \;=\; mg$$

$$w_e \;=\; mg - m\omega^2 R,$$

where R is the radius of the earth, and ω is the angular velocity of its rotation. Consider the ratio

$$\frac{w_e}{w_p} = 1 - \frac{\omega^2 R}{g}$$

The fractional loss f in weight is

$$f = \frac{\omega^2 R}{g}$$

Note that it is independent of the mass: this fractional loss applies to all masses.

To get a numerical answer everything will be expressed in the same units, feet and seconds.

ω is one rotation per day or

$$\omega = \frac{2\pi}{(24)(60)(60)} \frac{rad}{sec}$$

$$\omega = 7.272 \times 10^{-5} \, rad/sec$$

$$R = 4000 \, miles = (4000)(5280) \, feet$$

$$R = 2.112 \times 10^7 \, feet$$

$$g = 32 \, feet/sec^2$$

Using these values, the fractional loss in weight is

$$f = .0035 = .35\%$$

which is slightly greater than 1/3 of 1%.

5.5.4 Some Effects of the Coriolis Force on the Earth

The above result and the fact that the angular velocity of the earth is so small might lead one to believe that the effects of the rotation of the earth are on the verge of the unobservable and that the earth for all practical purposes can be considered as an inertial system.

This is not the case. There are a number of natural air and water movements whose circular configurations are often referred to as the Coriolis effect. They are all similar and include typhoons, hurricanes, tornados, whirlwinds, dirt devils and even whirlpools whether they be in the ocean, a lake or your bathtub.[23]

These phenomena are the result of a fictitious force that arises in the planetary system as a result of its rotational motion namely the Coriolis force. Imagine a gaseous or liquid mass on the surface of a planet. If such a mass moves either north or south, it has a velocity component in the direction of a radius from the axis of rotation. However, there is no provision on the surface to provide the mystery force to sustain movement along a radial line. If the mystery force were present, the mass would stay on (its radial) track and these phenomena would not occur. It is the absence of the mystery force that allows the fictitious Coriolis force to do its thing.

It almost seems like someone is pulling your leg when they say that these massive movements are caused by a force 'that isn't there' but nevertheless, the statement has content. Viewed from the accelerating system, the mystery

---

[23] It is important to note that the Coriolis effects require that opposite sides of the rotating mass be at different distances from the axis of rotation (of the earth or whatever). When this difference is small (as the whirlpool of your kitchen sink drain, or even a tornado) the tendencies are miniscule and can be dominated easily by other unrelated forces which may be in play. For larger circulations such as hurricanes, the Coriolis effects usually dominate the situation. In either case however, once these effects are established they tend to sustain themselves.

force would balance the fictitious Coriolis force and things could move radially without experiencing a net side thrust.

All the phenomena mentioned are caused in essentially the same way. The making of a fine tornado will be discussed as representative of this class of phenomena because it is the easiest to visualize from both an inertial system in space and from the non-inertial rotating earth.

Tornado

Ingredients: 1 low pressure area

Several high pressure areas strategically placed

1 planet with a gaseous atmosphere

To see how this works, refer to *figure* **5.18,** which shows the low pressure area in the northern hemisphere. First, imagine yourself in outer space in an inertial frame of reference viewing the earth as it rotates on its axis.

Initially, consider that the atmosphere is moving with the surface of the earth. (If this were not so, there would be a constant easterly wind whose velocity would be that of the surface of the earth at that latitude. At the equator, this is a little over 1000 miles per hour.)

First, consider the air mass moving up from the south. As it moves, it gets closer to the axis of rotation and is moving east with a velocity greater than that of the earth below. Hence, it veers to the east as shown. In contrast, the air from the north is getting farther away from the axis of rotation and therefore is moving too slowly to keep up with the surface below it. Hence, the surface moves eastward faster than the air and the air veers to the west. The unfilled vacuum of the low pressure area pulls these movements into a circular motion around itself providing the centripetal force necessary to sustain circular motion. The final result is a tornado consisting of a relatively low pressure core surrounded by a high-velocity circulation of winds.

A hurricane is essentially the same thing spread out over a larger area and less dramatic. The eye of the hurricane is the low pressure area in the center of the storm.

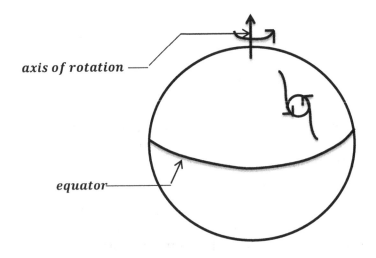

figure 5.18

Tornados, Hurricanes, Waterspouts, Dust Devils, Whirlpools, Etc

Note that all such phenomena produce counter-clockwise circulations in the northern hemisphere and clockwise circulations in the southern hemisphere. These phenomena do not occur near the equator because the movements of local air and water masses have too small a component in the directions toward or away from the axis of rotation of the earth. Near the equator, such movements are almost parallel to the axis.

To view this from the accelerating frame (the rotating earth), note that the actual real mystery force necessary to sustain side-thrust-free radial motion is $\vec{\omega} \times \vec{v}$. This force is absent from the scenario: there is nothing to slow down the winds from the south or speed up the winds from the north to keep them on a radial path. The fictitious Coriolis in accelerated frame is equal and opposite to the mystery force and hence can be written as $\vec{v} \times \vec{\omega}$ (recall that switching the order of the factors in a cross product changes the sign). For air moving in from the south $\vec{v}$ is toward the axis: when crossed into $\vec{\omega}$ which points up from the north pole, the resultant points east which agrees with the viewpoint from the inertial frame. Also the air moving in from the north has a velocity component away from the axis: when crossed into $\vec{\omega}$, the resultant points west, also in agreement with the viewpoint from the inertial frame.

5.6 POWER

5.6.1 Definition and Basic Formulas

Power is a very definite quantity in mechanics which corresponds to the less quantitative connotation of the word in ordinary language. We say that a machine or force is very powerful if it can deliver a great deal of work in a relatively short time. Another machine or force that can complete the same job but takes a longer time is said to be less powerful than the first.

Note that both machines (and forces) deliver the same amount of work. The concept of power therefore is not connected to the **amount** of work but rather to the **rate** at which it can be delivered. In keeping with this, we have the

...definition: power is the rate of doing work.

Translated into mathematics, the power **P** is defined by

$$P = \frac{dW}{dt}$$

where **W** is the work delivered.

Recall that the component of force normal to the direction of motion does no work. Hence the work done by a force depends on only its tangential component, and was defined by

$$W = \int F_s \, ds = \int f \cos(\theta) \, ds = \int \vec{F} * d\vec{s}0$$

A more basic approach would be to define the **differential** amount of work done by a force when it acts through a small (vector) distance $d\vec{s}$:

$$dW = \vec{F} * d\vec{s}$$

This is the equivalent of the integral definition except that this is one of the incremental contributions; the integral is their sum. If this last expression is divided by **dt**, the increment of time to travel the vector distance $d\vec{s}$, the result is a formula for the power:

$$P = \frac{dW}{dt} = \vec{F} * \frac{d\vec{s}}{dt}$$

$$P = \vec{F} * \vec{v}$$

where $\vec{v}$ is the velocity vector.

The basic effect of work done by a force is to change the speed of the object acted upon: the quantified version of this statement is that the work equals the change in kinetic energy. Hence, the differential work is equal to the differential change in kinetic energy:

$$dW \;=\; dT,$$

and once again, dividing by *dt*, we have

$$\frac{dW}{dt} \;=\; \frac{dT}{dt}$$

As a side point, notice that the formula for the power found above can be arrived at starting with the kinetic energy:

$$\frac{dW}{dt} \;=\; \frac{d}{dt}\left(\frac{1}{2}\,mv^2\right)$$

$$\frac{dW}{dt} \;=\; m\vec{v} * \frac{d\vec{v}}{dt}$$

$$\frac{dW}{dt} \;=\; v * \frac{d(mv)}{dt}$$

$$\frac{dW}{dt} \;=\; \vec{F} * \vec{v}$$

The formulas for power in terms of the angular parameters of motion are analogous. The differential work delivered by a torque is (see **section 4.5.2.**)

$$dW \;=\; -\;\tau\,d\theta$$

The power formula follows:

$$P \;=\; \frac{dW}{dt} \;=\; \tau\,\omega.$$

This is the analogue of $\vec{F} * \vec{v}$. We are not concerned with the dot product version because for purposes here, the torque and the angular velocity are always in the same direction.

5.6.2 Some Examples of Power Calculations

Free Fall in the Local Gravity Field:

The local gravity force on a mass m is $-mg$: if a body falls vertically in this field, its velocity is $-gt$ (assuming it started at zero velocity). Then the power delivered is

$$P = \vec{F} * \vec{v}$$

$$P = mg^2 t$$

Notice that even though the force is constant the power delivered to the mass is not.

To verify that this is equal to the time derivative of the kinetic energy T, write T as a function of time:

$$T = \frac{1}{2} mv^2$$

$$T = \frac{1}{2} mg^2 t^2$$

$$P = \frac{dT}{dt} = mg^2 t$$

which agrees with the previous result

.

Power in a Spring-Mass System:

As a second example, consider the motion of a spring-mass system

$$x = A \sin(\omega t).$$

The natural angular velocity is

$$\omega = \sqrt{\frac{k}{m}}.$$

410

The power can be computed from $\vec{F} * \vec{v}$:

$$F = -kx$$

$$F = -kA\sin(\omega t)$$

$$v = \frac{dx}{dt}$$

$$v = A\omega\cos(\omega t)$$

$$P = Fv$$

$$P = -kA^2\omega\sin(\omega t)\cos(\omega t)$$

$$P = \frac{-kA^2\omega}{2}\sin(2\omega t).$$

The last form illustrates a point: the power swings positive and negative in a symmetric fashion. This indicates that the spring delivers power to the system and moves the mass, then it absorbs power from the system and the mass stops, then it delivers it again, and so on.

This makes sense because the spring-mass system while conserving total energy, exhibits an energy flow back and forth between kinetic and potential. The power relates only to the kinetic energy and hence refers to only this half of the exchange.

Finally, it will be verified that the power can also be determined as the time derivative of the kinetic energy:

$$T = \frac{1}{2}mv^2$$

$$T = \frac{1}{2}mA^2\omega^2\sin(\omega t)\cos(\omega t)$$

If the fact that $\omega^2 = k/m$ is used the previous expression reduces to that obtained above.

The power is a quantity of interest when one wishes to design a mechanical system capable of performing a task to given specifications or when one wishes to determine the energy flow from one part of a system to another. It is usually not possible to determine the power until the rest of the quantities associated with the motion are known. In most situations therefore, the motion of the system is known before one computes the power. It follows that it is generally computed by direct application of the formulas as in the above examples.

CHAPTER VI

The Universal Gravitational Field

6.1 THE UNIVERSALITY OF GRAVITATIONAL ATTRACTION

6.1.1 The Content of this Chapter

The large portion of this chapter is devoted to the solution of the two body problem, i.e. the derivation of the motion exhibited by two bodies orbiting around each other as a result of their mutual gravitational attraction. The analysis of the problem is somewhat of a project and is therefore broken down into three sub-analyses. Before embarking on the project however, there is a short description of Newton's law of universal gravitation which he found to be in agreement with his law of the dynamics of mechanical systems. Following that is a discussion of the pre-Newtonian investigations that laid some of the groundwork, namely, Kepler's three laws of planetary motion. There is an elementary discussion of these laws in relation to circular orbits which are a special case of the more general elliptical ones proclaimed as a portion of Kepler's observations. It will be shown that his third law fixes the gravitational force as an inverse square law even for the special case of circular orbits.

The project of finding the solution to the two-body problem follows. It is broken down into three sub-problems. The first of these is referred to as the reduction to a one-body problem. It is shown that the motion of one of the bodies satisfies the inverse square law of gravitation around the *CM* if the mass of the other body is appropriately scaled and considered to be at rest and located at the **CM.** The second sub-problem is mathematical in nature: it consists of a detailed analysis of ellipses. The comparison of these results with the solution of the differential equation for the orbit relates the physical quantities involved with the parameters of the ellipse. The actual solution of the orbital equation is the third sub-problem.

The remainder of this chapter addresses issues of gravitational field theory. It includes the determination of the fields for a few simple geometric configurations of field sources. In this regard, the field sources will be finite masses **m** or infinitesmal **dm's**. The field resulting from any configuration of mass-sources has an analogous electro-static field for a similar configuration of charge-sources (finite **q's** or infinitesmal **dq's**). This similarity follows from the fact that the gravitational law is

$$F \quad = \quad G\frac{m_1 m_2}{r^2}0$$

And the electro-static law is

$$E \quad = \quad K\frac{q_1 q_2}{r^2}.0$$

In each of these cases, the force field varies inversely as the square of the distance from a point source. The constants are different, but for a given source configuration, this is mathematically non-essential to the resulting field configuration.

Following the calculation of several field configurations is a lengthy description of the flux of a vector field across a surface. The more general flux equations are then particularized to a discussion of Gauss' law which for inverse square fields relates the flux across an arbitrary closed surface to the field sources enclosed by the surface. In particular, it relates the flux of the gravitational field across an arbitrary surface to the mass enclosed by the surface. At the end of the chapter is the derivation of the differential form of Gauss' law which is generally used to determine the configurations of inverse square fields.

For the following, P represents a typical but unspecified point (x, y, z) in space. Three methods for determining the resultant field configurations will be illustrated:

1. *find the contribution to the (vector) field at P from one typical source-point and add (vectorially) the contributions from all such source-points to find the net field at P;*
2. *find the contribution to the potential at P from one typical source-point and add the contributions from all such source-points to find the net potential at P. Find the field at P by taking the three principal directional derivatives at P;*
3. *use Gauss' law which involves the flux of the unknown field over a closed surface. If by symmetry, it can be concluded that the field is constant over some surface, and that surface is chosen to apply the theorem, then the unknown field can be taken out of the integral and the field can then be found from the geometry (this will be clarified further when we get to it).*

Some of the source-configurations are unlikely as mass distributions but nevertheless, represent reasonable charge distributions. The field configurations will be found anyway. Keep in mind that in these analyses, it is the methodology that is important, not the particular problems considered.

Chapter VI: Universal Gravitation

6.1.2   The Law of Gravitation

The law of universal gravitational attraction was first announced by Newton in the year 1686. In essence, he said

every particle of mass in the universe attracts every other one with a force that is proportional directly to the product of their masses and inversely to the square of the distance between them.

In symbols this is written

$$F \quad \propto \quad \frac{m_1 m_2}{r^2}$$

This is a statement of the proportionality proclaimed above. It can be made into an equation by inserting the proportionality constant G known as the universal gravitational constant:

$$F \quad = \quad G \frac{m_1 m_2}{r^2}$$

It is not known exactly how the universality of this law occurred to Newton, but it is believed that it was in connection with the observation and study of planetary motions: in particular, that of the moon around the earth. The groundwork for this work was laid by Kepler whose observations led experimentally to three conclusions known as Kepler's laws.

In the late 1700's Coulomb designed an extremely sensitive and simple device for the measurement of electrostatic forces of attraction and repulsion. In 1798, Cavendish adapted the design to measure the universal gravitational constant G. The device consists of a torsion pendulum made from a thin quartz thread from which are suspended two identical masses connected by a light rod as shown in *figure* **6.1**. A light beam is reflected off a mirror which is attached to the thread onto a calibrated scale which is set to zero at equilibrium.

When two identical masses are brought into the proximity of the suspended masses, their gravitational attraction to the suspended masses produces an angular deflection of the pendulum which is proportional to the force they exert. The deflection is magnified by the mirror arrangement allowing for the measurement of very small forces.

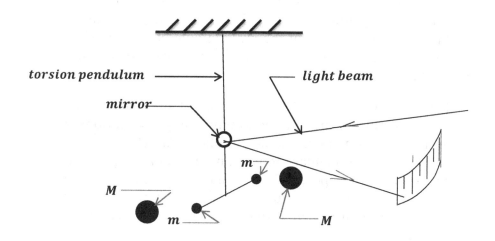

figure **6.1**

Cavendish's Adaptation of Coulomb's Apparatus

The gravitational constant is

$$G = 6.67 \times 10^{-8} \ \frac{dyne - cm^2}{gm^2} \ in \ cgs \ units$$

$$G = 6.67 \times 10^{-11} \ \frac{newton - m^2}{kgm^2} \ in \ mks \ units$$

6.1.3 Kepler's Laws

On the basis of his observation of the solar system, Kepler drew three experimentally determined conclusions concerning the motion of the planets.

These results presented around 1610 provided the experimental conclusions which agree with Newton's dynamics equation $\vec{F} = m\vec{a}$. The third law, as will be shown, fixes the form of the gravitational attraction as an inverse square law, even in the special case of circular orbits.

The second Keplerian law falls short of the mark in the determination of the laws governing planetary motion: in terms of Newtonian mechanics, it follows directly from the conservation of angular momentum around a single point and hence is a property of any radially directed force field whether it be inverse-square or not. The first and third laws however, are peculiar to inverse square attraction.

6.1.4 Kepler's Laws for Circular Orbits

Geometrically, the circle is a special case of an ellipse. It would therefore seem that certain initial conditions would result in a circular orbit. Although this is about to be shown, it should be kept in mind that circular orbits are possible for any attractive force field that is centrally directed. It is the fact that the other non-specifically chosen orbits are ellipses that conjoins with Kepler's first law to determine the form of the gravitational law. However, as mentioned above the tlaw single-handedly makes this determination even for the specially chosen circular orbits.

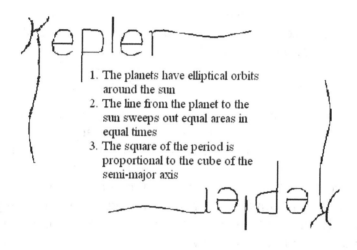

1. The planets have elliptical orbits around the sun
2. The line from the planet to the sun sweeps out equal areas in equal times
3. The square of the period is proportional to the cube of the semi-major axis

figure 6.2

A Tribute to Johannes Kepler

In general, the force of attraction between the two bodies could depend on its angular position as well as the distance between them. But since there is no a priori reason to believe that there should be a preferred direction in empty space, it is reasonable to assume that the force of attraction depends only on the distance between the bodies and is radially directed. In lieu of these remarks, the force of attraction will be written here as

$F(r)$, using the single variable r in the argument to indicate that it does not depend on the angular position.

The condition for a circular orbit is that the attractive force be equal and opposite to the centrifugal force:

$$F(r) \ = \ -m\omega^2 r$$

$$\frac{F(r)}{r} \ = \ -m\omega^2$$

where r is the distance between the bodies. For a circular orbit, r is a constant and consequently the left side of the above equation is also. Thus r determines a particular angular velocity for which a constant-speed circular orbit is possible. This works for any force of attraction $F(r)$.

That equal areas are swept out in equal times is now obvious since the angular velocity is constant. But this is true for any path of motion determined by a centrally directed force between two bodies: it follows directly from the conservation of angular momentum. (Angular momentum is conserved around the center point of any centrally directed force field because there can be no torque around this point.) We have

$$L \ = \ mr^2 \frac{d\theta}{dt}$$

The diagram *in figure 6.3* shows a general orbit indicating the differential area dA swept out in a differential time dt as the angular position changes by the angle $d\theta$. The area of the triangle swept out is

$$dA \ = \ \frac{1}{2} r^2 d\theta \, .$$

$$\frac{dA}{dt} \ = \ \frac{1}{2} r^2 \frac{d\theta}{dt} \ = \ \frac{L}{2m} \ = \ constant.$$

Hence, the rate of change of area is a constant which is just another way of saying that equal areas are swept out in equal times. Once again this applies to any centrally directed attraction (or repulsion) and does not (for any shaped orbit) imply an inverse square law attraction.

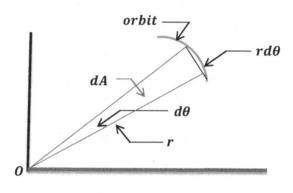

figure 6.3

Centrally Directed Field

When applied to a circular orbit, Kepler's third law has a stronger implication than the first two. Note first that the angular displacement θ is

$$\theta \;=\; \omega t.$$

When the time is one period T, the body has made one complete revolution. Hence, θ at that time is 2π. In other words, if T is the period

$$2\pi \;=\; \omega T.$$

To relate the period to the radius r of the orbit (for a circular orbit, the semi-major axis is just the radius) start with the condition necessary to maintain a circular orbit:

$$F(r) \ = \ -m\omega^2 r \, ;$$

Multiply both sides of the equation by T^2 and compare with the above equation for T;

$$F(r)T^2 \ = \ -m(\omega T)^2 r \, .$$

Noting that the square on the right side equals $4\pi^2$:

$$F(r)T^2 \ = \ -4\pi^2 mr \, .$$

But Kepler's third law states that the square of the period is proportional to the cube of the semi-major axis: $T^2 = kr^3$ where k is the proportionality constant. Substituting for T^2 in the previous equation,

$$F(r)kr^3 \ = \ 4\pi^2 mr$$

$$\Rightarrow \quad F(r) \ = \ \frac{4\pi^2 m}{kr^2}$$

The above analysis indicates that Kepler's third law can be satisfied only if the gravitational attraction varies inversely as the square of the distance r. This fixes the form of the gravitational law of attraction as an inverse square law.

6.2 THE TWO BODY PROBLEM

6.2.1 The Reduction to a One Body Problem

It was remarked above that the general two body problem could be referred to the **CM** system. When this is done, the distance of the masses from the origin is no longer the distance (between the masses) which occurs in the gravitational law. Thus the form of the radially directed force comes into question when viewed from the *CM* system.

It will now be shown that it is possible to consider the motion of one of the masses around the *CM* and that the form of the gravitational force equation remains intact if:

1. *the second mass is fixed at the CM (taken as the origin);*
2. *the second mass is appropriately scaled in value.*

In this way the original problem is replaced by a slightly different one which is considerably simpler without affecting the motion of the first mass.

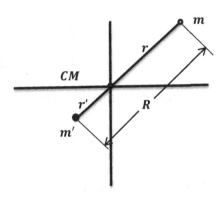

figure 6.4

Gravitational Law in a CM System

The system is shown in the diagram of *figure 6.4* with the *CM* at the origin. Let the masses be *m* and *m'* at respective distances r and r' from the CM. Let R be the distance between the bodies. Note that there are no outside forces on the system. Therefore, any coördinate system with its origin at the CM is an inertial system and can be considered stationary without any loss of generality.

Although the following seems like an algebraic slight-of-hand trick, it accomplishes the simplification (promised earlier) that accompanies the use of a CM system to solve problems. Fill in the missing algebra steps as we go along.

Proceed by relating the distances and the masses. The moments around the CM must be zero. Therefore,

$$mr \;=\; m'r'$$

Also the sum of the distances r and r' is R:

$$r + r' \;=\; R$$

$$\Rightarrow \quad r \;=\; \frac{m'R}{m + m'}$$

$$r' \;=\; \frac{mR}{m + m'}$$

Recall that the force of attraction between the two masses is

$$F \;=\; G\frac{m'm}{R^2}$$

The mass m will now be singled out and its position will be referred to the CM system. This requires that the gravitational equation be rewritten in terms of r:

$$R \;=\; r + r'$$

$$R \;=\; r\left(1 + \frac{m}{m'}\right)$$

$$\Rightarrow \quad r \;=\; \left(\frac{m'}{m + m'}\right)R.$$

Let μ be the constant defined by

$$\mu = \left(\frac{m'}{m + m'}\right)$$

and multiply the numerator and denominator of the force equation by μ^2 :

$$F = G\frac{(m'\mu^2)m}{(\mu R)^2} ,$$

$$F = G\frac{Mm}{r^2} ,$$

where $M = \mu^2 m'$. This is the original force equation with some substitutions. However, this last form indicates that the motion of the mass m around the CM (taken at the origin) can be found from the above equation if

1. *the second mass m' is replaced by $M = \mu^2 m'$;*
2. *M is positioned at the origin where it is fixed;*
3. *r is the distance from the origin to the mass m.*

Thus, a two-body problem can be reduced to a one-body problem without loss of generality. It is the last equation that will be solved under the three conditions cited above. Notice that once again, a considerable simplification has been accomplished by considering the problem in a CM system.

It is expected that the solution to the last equation be an ellipse ala Kepler and it is therefore expedient at this time to gather some pertinent information. So, "we interrupt this program to bring you ...

6.2.2 The Analytic Geometry of the Ellipse"

6.2.2.1 General Remarks

It is time to go play in the ellipses for a while to develop some sense of what an ellipse is and to define the parameters used to describe them.

There are four basic parameters: the semi-major axis; the semi minor axis; the eccentricity; the distance from the center to one of the foci. Only two of these are independent and it is necessary to know all the relations between them.

Basically, there are two steps in the following: obtain the equation for the ellipse when the origin is at its center; transform this equation to that of the same ellipse when the origin is at one of the foci. The second step is necessary because the origin of the *CM* system in the two-body problem will be at one of the foci when the problem is finally solved.

Eventually, the equation for an ellipse will be determined from its locus definition, but before this is approached, it is remarked that an ellipse is simply a circle which is distorted in two orthogonal directions, say x and y, by simple scaling factors. Start with the unit circle:

$$X^2 + Y^2 = 1$$

and make the substitutions

$$x = aX;$$
$$y = bY.$$

Thus X is scaled by a factor of a and Y, by a factor of b. Writing the equation in terms of x and y,

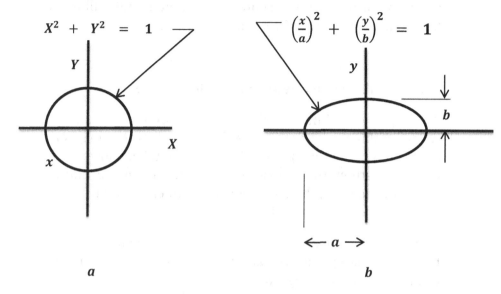

figure 6.5

Distorting a Circle

In terms of X and Y, this is still a unit circle, but in terms of x and y, it is distorted by the scaling factors and the result is an ellipse.

Both terms on the left are positive: hence, when $x = \pm a$, $y = 0$ and when $y = \pm b$, $x = 0$. (See *figure 6.5*.) The larger of the two numbers a and b is called the **semi-major axis** (i.e. the biggest 'radius') and the smaller is called the **semi-minor axis** (i.e. the smallest 'radius'). To fix ideas here, 'a' will always be the semi-major axis and will always lie along the x-axis.

The last equation is therefore the equation for an ellipse when the origin is at the center. This same result will now be obtained starting from the locus definition.[24]

6.2.2.2   The Locus Definition of the Ellipse

An ellipse is defined as the locus of points the sum of whose distances from two fixed points called the foci, is a constant. If these two points (the foci) are taken closer and closer together, the ellipse looks more and more like a circle. In the limit, the foci coïncide with each other at the center of the figure, and the curve is a circle.

Figure 6.6 shows the foci f and f' on opposite sides of the origin along the x-axis, each a distance c from the origin. With this configuration, the semi-major axis will be along the x-axis and in keeping with the locus definition, the sum of the distances r and r' from each of the foci to an arbitrary point P on the curve is a constant. If the point R where the curve crosses the major axis is considered, it is seen that the sum of these distances is $2a$, the major axis ('a' is the **semi**-major axis).

Now consider the point Q where the curve crosses the y-axis: the sum of the distances from the foci to Q is $2a$; hence, each focus is a distance 'a' from the tip of the semi-minor axis b and we have

$$a^2 \ = \ b^2 \ + \ c^2$$

---

[24] If a cone, extended on both sides of its apex is sliced by a plane, the intersection can be any one of four figures: a circle, an ellipse, a parabola or a hyperbola. The particular curve you get depends on the relative orientations of the cone and the plane. These four figures are called the conic sections and their importance in the field of optics has motivated their extensive study. Each of them has a locus definition from which its equation can be determined in much the same way as that presented here for the ellipse.

which relates the semi-axes to c, the distance from the center of the ellipse to one of the foci.

The ratio c/a is a measure of how much the ellipse deviates from a circle. It is an important parameter and is given a special name: it is called the ***eccentricity*** of the ellipse and is designated ε:

$$\varepsilon \;=\; \frac{c}{a}.$$

It follows that

$$b \;=\; a\sqrt{1 - \varepsilon^2}$$

Note that when $\varepsilon = 0$, the figure is a circle; when $\varepsilon = 1$, it is a straight line. For any ellipse, the eccentricity must be in the range $(0, 1)$.

The locus definition requires that

$$r + r' \;=\; 2a$$

$$\sqrt{(x - c)^2 + y^2} + \sqrt{(x + c)^2 + y^2} \;=\; 2a$$

Note that when $\varepsilon = 0$, the figure is a circle; when $\varepsilon = 1$, it is a straight line. For any ellipse, the eccentricity must be in the range $(0, 1)$.

The locus definition requires that

$$r + r' \;=\; 2a$$

$$\sqrt{(x - c)^2 + y^2} + \sqrt{(x + c)^2 + y^2} \;=\; 2a$$

Let A be the quantity under the first radical and B, the quantity under the second. Then we have

$$\sqrt{A} + \sqrt{B} \;=\; 2a$$

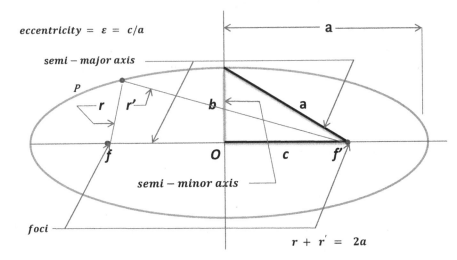

figure 6.6

The Mathematics of the Ellipse

The equation will have to be squared twice to clear the radicals. Squaring it once,

$$A + 2\sqrt{AB} + B = 4a^2$$

$$2\sqrt{AB} = 4a^2 - (A + B)$$

Squaring it again,

$$4AB = 16a^4 - 8a^2(A + B) + (A + B)^2$$

Substitute the expressions for *A* and *B*:

$$0 = 16a^4 - 16a^2(x^2 + c^2 + y^2) + (-4cx)^2$$

$$(a^2 - c^2)x^2 + a^2y^2 = a^2b^2$$

$$b^2x^2 + a^2y^2 = a^2b^2$$

$$\left(\frac{x}{a}\right)^2 + \left(\frac{y}{b}\right)^2 = 1$$

This is the equation of the ellipse when the origin is at the center (at the midpoint of the line joining the foci).

6.2.2.3   The Equation of the Ellipse When the Origin Is at a Focus

Refer to *figure 6.6*. In preparation for solving the equations for the two-body problem, we will now place the origin at one of the foci (in particular the left one at *(x = −c)*, and write the equation of the ellipse in terms of the angular coördinates *r* and *θ*. (These are usually referred to as polar coördinates.) In this part of the analysis, the parameters *a* and *ε* will be used. The semi-minor axis *b* will be eliminated using the relation

$$b^2 = a^2(1 - \varepsilon^2)$$

Then the above equation for the ellipse becomes

$$(1 - \varepsilon^2)\left(\frac{x}{a}\right)^2 + \left(\frac{y}{a}\right)^2 = (1 - \varepsilon^2)$$

The substitutions necessary to change the variables to *r* and *θ* are (refer to *figure 6.6*):

$$x = r\cos(\theta) - c$$
$$y = r\sin(\theta)$$

Also substitute $R = r/a$ to simplify the algebra a little and the above equation becomes

$$(1 - \varepsilon^2)[R\cos(\theta) - \varepsilon]^2 + [R\sin(\theta)]^2 = 1 - \varepsilon^2$$

$$R^2\cos^2(\theta) - 2\varepsilon R\cos(\theta) + \varepsilon^2 - \varepsilon^2 R^2\cos^2(\theta)$$

$$+2\varepsilon^3 R cos(\theta) - \varepsilon^4 + R^2 sin^2(\theta) = 1 - \varepsilon^2$$

$$R^2 - \left[\varepsilon^2 R^2 cos^2(\theta) + 2\varepsilon R(1 - \varepsilon^2)co\, s(\theta) - (1 - \varepsilon^2)^2\right] = 0$$

$$R^2 \quad - \quad [\varepsilon R cos(\theta) + (1 - \varepsilon^2)]^2 \quad = \quad 0$$

The last expression is the difference of two squares, which can be factored:

$$R \quad - \quad [\varepsilon R\, cos(\theta) + (1 - \varepsilon^2)] \quad = \quad 0$$
$$R \; + \; [\varepsilon R\, cos(\theta) + (1 - \varepsilon^2)] \quad = \quad 0$$

The second of these equations leads to a negative value of r. Although it is correct, it is preferable to take $r > 0$ whenever possible as negative values of r indicate a **180°** shift in $\boldsymbol{\theta}$ which is troublesome. We therefore use the first solution with the result

$$r \quad = \quad \frac{a(1 - \varepsilon^2)}{1 - \varepsilon cos(\theta)}$$

This last expression is the equation for the ellipse when the origin is at the focus to the left.

Incidentally, this equation is often written in its reciprocal form

$$\frac{1}{r} \quad = \quad k[1 - \varepsilon\, cos(\theta)],$$

where

$$k \quad = \quad \frac{1}{a(1 - \varepsilon^2)}.$$

6.2.3 The Orbital Equation and its Solution

Since the force is radially directed, the distance r and the angle θ constitute a more natural coördinate system in which to describe the problem. Recall that the equations of motion involve the second derivatives of r and θ. Their solution consists of two functions: the distance $r(t)$ as a function of time and the angle $\theta(t)$ as a function of time. These two functions are a complete solution to the problem.

If one were to get the complete solution, the time t could be eliminated from $r(t)$ and $\theta(t)$ resulting in a single function $r(\theta)$: this function is the equation for the orbit.

The process of finding the complete solution as described above requires methods for solving differential equations which are too far afield from the present subject matter. Therefore, the functions $r(t)$ and $\theta(t)$ will not be found. We will however, obtain the orbital equation by eliminating the time dependence. The differential equation $r(\theta)$ for the orbit will then be found and this equation will be solved.

The following is the program to be followed:

1. *write the equation of motion for the angular direction. It is a second order differential equation but can be integrated once immediately. The information it affords is that the angular momentum is conserved;*

2. *instead of the radial equation of motion, the conservation of energy will be used. The result is a first order equation;*

3. *these two results will then be used to eliminate the time parameter before solving them. The result will be a first order differential equation for the orbit (i.e. for r as a function of θ rather than as a function of time);*

4. *the solution of the orbital equation will then be accomplished without the use of methods for solving differential equations. This happy situation is possible because the equation can be integrated directly.*

Do not beat yourself up with the belief that you should at this stage of the game be able to concoct this solution or find an equivalent one. You should be able to follow the steps or at least almost follow them. Be aware that the equations are non-linear and the solution has been obtained through a good deal of educated futzing around. This situation concerning the solution of non-linear equations was discussed earlier in *section 3.2.3* in connection with the non-linear equations for perfectly elastic collisions.

Furthermore, this problem which is the simplest one encountered in celestial mechanics has been extensively studied for a long time and has been whittled down to essentials by a number of contributing workers. (You are presently in the midst of one such whittled-down version.)

Step 1

The equation of motion for the angular direction is

$$\sum \tau = \frac{dL}{dt}$$

$$0 = \frac{d(mr^2\omega)}{dt}$$

The torque is zero because the force is centrally directed. Hence, the equation can be integrated:

$$L = mr^2\omega$$

where the angular momentum L is the constant of integration. We could have started at this point because we already knew that for centrally directed forces, angular momentum around the center is conserved. This relation will now be used to eliminate ω wherever it occurs:

$$\omega = \frac{d\theta}{dt} = \frac{L}{mr^2}.$$

Step 2

In *section* **1.3.5.3** it was shown that a centrally directed force field whose force depended only on the radial distance r from the central point, was conservative. The gravitational field was considered in connection with this and it was shown that its potential energy was given by

$$V = -G\frac{Mm}{r}.$$

We will now use the conservation of energy. The total energy H is a constant of the motion and is given by

$$H = V + \frac{1}{2}mv^2.$$

Step 3

To relate the energy H to r and θ, we write the velocity in terms of its radial and angular components:

$$v_r = \frac{dr}{dt}$$

$$v_\theta = r\omega = r\frac{d\theta}{dt}$$

$$v_\theta = \frac{L}{mr}$$

where the time derivative was eliminated using the result of step 1.

The square of the speed is

$$v^2 = v_r^2 + v_\theta^2$$

Collecting these items, the total energy H is

$$H = -\frac{GMm}{r} + \frac{1}{2}m\left[\left(\frac{dr}{dt}\right)^2 + \frac{L^2}{mr^2}\right]$$

The remaining time derivative is now eliminated using the chain rule:

$$\frac{dr}{dt} = \frac{dr}{d\theta}\frac{d\theta}{dt} = \frac{L}{mr^2}\frac{dr}{d\theta}.$$

Using this in the above equation, the final expression for the total energy is

$$H = -\frac{GMm}{r} + \frac{L^2}{2m}\left[\frac{1}{r^4}\left(\frac{dr}{d\theta}\right)^2 + \frac{1}{r^2}\right].$$

Solving for $dr/d\theta$,

$$\frac{dr}{d\theta} = \sqrt{\frac{2mH}{L^2} + \frac{2GMm^2}{L^2}r^3 - r^2}$$

To facilitate the evaluation of the final integral, the above equation is put in the form

$$d\theta = \frac{\frac{dr}{r^3}}{\sqrt{\frac{2mH}{L^2} + \frac{2GMm^2}{L^2}\frac{1}{r} - \frac{1}{r^2}}}.$$

Step 4

Notice that most of the complicated looking gibberish in the last equation consists of constants. This is often the case in physics and if you want to relate the energy, angular momentum, gravity constant, etc. to the parameters of the motion, r and θ in this case, they have to be carried along. Don't let the algebra laden with all these constants mask the essence of the mathematical form: the form of the above equation is

$$d\theta = \frac{dr}{r^2\sqrt{A + \frac{B}{r} - \frac{1}{r^2}}}$$

where the constants A and B are defined by an obvious comparison of this expression with the previous one.

The integral can be solved by elementary methods: transform the equation by making the substitution

$$u = \frac{1}{r}$$

$$\Rightarrow \quad du = -\frac{dr}{r^2}$$

Then the above equation can be written in terms of u:

$$d\theta = \frac{-du}{\sqrt{A + Bu - u^2}}$$

Complete the square under the radical (i.e. add and subtract $B^2/4$):

$$d\theta = \frac{-du}{\sqrt{A + \left(\frac{B}{2}\right)^2 - (u - \frac{B}{2})^2}}$$

After some algebra, this can be written

$$d\theta = \frac{-d\left[\dfrac{u - B/2}{\sqrt{A + (B/2)^2}}\right]}{\sqrt{1 - \dfrac{(u - B/2)^2}{A + (B/2)^2}}}$$

whose form is

$$d\theta = \frac{-dw}{\sqrt{1 - w^2}}$$

with

$$w = \frac{u - B/2}{\sqrt{A + (B/2)^2}}$$

In the last expression for $d\theta$, the right side of the equation is the differential for the arc-cosine of w. The integral is

$$\theta + \theta' = \cos^{-1}(w).$$

The constant of integration, θ', rotates the angle θ. If it is taken as zero, then the semi-major axis coïncides with the x-axis and

$$w = \cos(\theta),$$

Substituting the expression for w, and straightening out some of the algebra, you get

$$\frac{1}{r} = \frac{B}{2} + \sqrt{A + \left(\frac{B}{2}\right)^2} \cos(\theta).$$

Referring back to the end of *section 6.2.2.3*, this result is seen to be the equation for an ellipse (in its reciprocal form) when the origin is located at one of the foci. The mass *m* therefore moves in an elliptical path around the *CM* of the system.

Exercise

Substitute the expressions for *A* and *B* in the final answer above to relate the orbital parameters to the parameters of the ellipse (i.e. compare your result with the equation for the ellipse obtained at the end of **section 6.2.2.3.**)

This problem is the first and simplest one encountered in celestial mechanics and as such is deserving of some detailed study.

6.3   FIELD THEORY

6.3.1   The Concept of a Force Field

In the preceding analysis of the two-body problem, it was tacitly assumed that the bodies were point masses. In what follows, the gravitational field for extended bodies will be considered and it will be shown that outside the mass, the field for a spherical source is the same as it would be if all its mass were concentrated at its center. This will justify the analysis of the two body problem which proceeded as though the planets were point masses.

Before investigating the details of the gravitational field, we will ascertain just what is meant by a force field. Note that in the context of gravitation, what is observable is the force that exists when there are two masses present. The following hypothetical experiment allows for a detailed description of the process by which we proceed from observable forces to non-observable fields.

Consider two masses, *m* and *m'*: *m* will be left unmolested during the process; *m'* will be used as a test mass and will be allowed to vary. When *m'* is introduced into the region proximate to *m*, it is subject to an attractive force which is measured as

$$F \;\; = \;\; -G\frac{mm'}{r^2}$$

where *r* is the distance between the (centers of the) masses. Now, if *m'* is moved to another location with respect to *m*, a force different in both magnitude and direction is observed and found still to be in accordance with the above equation. Then at another location, then another, etc. It can be construed from the fact that the force vectors can be determined from the above equation and that no force is observed when the mass *m* is removed, that *m* is the **cause** of these forces and that its existence somehow predisposes the space in its vicinity.

If we are to pursue this idea, we are faced with the task of quantifying the preconditioning of the space by *m* and it must be quantified in such a way that it is independent of *m'*. We note first that *F* depends on the size of *m'*. We attempt to eliminate this aspect of the problem by considering *F'*, the force per unit *m'*:

$$F' \;\; = \;\; \frac{F}{m'}.$$

Next we argue that the very presence of *m'* may be altering the condition of the local space, so we try to minimize any possible effect by letting *m'* become smaller and smaller, presuming that if it is sufficiently small, it can't be screwing things up too much. Mathematically, this is taking the limit as $m' \to 0$.

We conclude at this point that what we have is something which depends only on m and since this 'something' extends throughout the region of space near m we call it a field, and since it is associated with the force of gravity, we call it the **gravitational field produced by the mass m.**

Let U represent this gravitational field produced by m. Then the above description in mathematical symbols is

$$U \;=\; \lim_{m' \to 0} F' \;=\; \lim_{m' \to 0} \frac{F}{m'}$$

$$U \;=\; -G\frac{m}{r^2} \lim_{m' \to 0} \frac{m'}{m'}$$

$$U \;=\; -G\frac{m}{r^2}.$$

This was the argument intended to justify the existence of the electrostatic field and although over a century late, it was also used to justify the existence of the gravitational field.

The field U quantifies the presumed preconditioning of the space in the vicinity of m. It is contrived in such a way that if an arbitrary second mass m' is introduced into the region, then the force acting on it is

$$F \;=\; m'U$$

Do you buy it?? ..... Well don't!

6.3.2   The Emperor's New Clothes

With pomp and ceremony did the emperor parade naked through the streets of the town, and everyone believed himself unworthy to see his magical clothes which, they were told, would be visible only to those who were without blemish. Then a child in all innocence turned to his mother and said, "The emperor has no clothes!" and then ...

Classical field theory parades similarly. Its tailors have dressed it in a logic that isn't there, embellished it with mathematical finery which does not fit and cloaked it in a robe of mystery behind which it simply cannot hide. Let's examine the definition of the previous section a little more closely.

The issue of logic enters at the beginning of the section with the statement, "...*can* be construed ... that *m* is the cause of these forces...". The fact that an existing condition *can* be construed as implying the existence of a field is not compelling enough logically to instate such a field as a physical reality. A metaphysical conclusion about existence or non-existence requires more than construing. A lengthy discussion of this point was presented in *section* **3.3.1**.

A second issue concerning the logic is the limiting process which is used to define the classical field. Its conclusion follows from a premise that pretends to be logical but isn't: namely, it assumes that *'almost not there'* is *almost* the same as *'not there'.* Indeed, the concept of 'almost not there' in itself is a little strange, if not downright comical. But, delivered with the proper pomp and ceremony, one could easily believe himself unworthy to comprehend the mystery at hand.

It should also be noted that the gravitational force is related to m' as a simple proportion. When the force was divided by m', to find the force per unit m', the result was already independent of m'. The subsequent limiting process is, in the case in point, totally ineffectual. It stands as a pseudo-philosophical argument to convince us that the field is really there. In the final analysis, we have simply divided out the troublesome m'; any subsequent embellishment such as the limiting process attempts to justify and/or mask this embarrassing fact.

A third objection issues from the fact that matter is not continuous, it is atomic. It is questionable whether or not a continuous limiting process is a suitable mathematical description. It might be argued that the calculus of continuous variables is used extensively throughout physics and that this is done with full knowledge that this description is an approximation and is valid only within certain limits. But the situation here is quite different: we are not analyzing a system which lends itself to 'approximate continuity' (whatever that could be); we are philosophizing about the existence or non-existence of a physical entity. As in the case of the first objection, this demands more logical stringency.

So indeed, the emperor has no clothes...but...

... he is still the emperor. The tailor and those in attendance who raved about his clothes may have unresolved business with the emperor and the emperor himself may have some gullibility issues, but he still reigns. And so it is with fields.

It is not the point here to discredit fields and the immensity of their power in the description of physical situations; the purpose is to instill a clear understanding of their relationship to physics. The definition of classical fields establishes them as an extremely useful mathematical device but not as physical realities. It is of great importance to make use of these alleged fields in the analyses of physical systems; it is equally important to know that they have not been established as realities. (See *section* 3.3.1.1.)

Neither quantum theory nor relativity can sustain the concept of a classical field. In quantum mechanics, only potentialities exist until made manifest in one form or another *by observation.* Hence, reality is much more severely restricted to observability in quantum mechanics than in classical physics. Relativity removes part of the objection by removing the 'pre' in 'predisposed': in relativity the space is conditioned (disposed) by all the masses present at the time of their interaction; it cannot be conditioned by some of them before time. Furthermore, the disposition of the space in general relativity, i.e. its curvature, is not a theoretically unmeasurable quantity. (See *section* 3.3.1.4.)

It should be understood that the concept of a classical field is offensive to modern physics. It is here that the use of fields requires that one be very clear about their status. In classical physics however, one can virtually treat fields as a reality: here things seem to act *as if* the fields were really there and they do so pretty universally within the confines of the classical context.

It should be emphasized that problems in classical field theory are not at all limited to classical times; for example, there are many problems in modern plasma physics which are essentially problems in classical field theory.

Incidentally, if it is recalled that the classical gravitational force and the classical electrostatic force have the same form (inverse square law) and are justified in the same way, then at least in this context, it is generally believed that what is said of one applies also to the other. The only known reasonable approach to many problems that occur even today is the use of classical field theory which when considered in the context of all the scientific endeavors of mankind, has increased in importance over the years even though fields are no longer considered to be physical realities. In the final analysis, field theory is no more dispensable today than is $F = ma$.

6.3.3 Mathematical Methods for Finding Fields

6.3.3.1 General Description

The classical theory of gravitation starts with the assumption that the field intensity at a distance r from a point mass m is

$$U = -G\frac{m}{r^2}.$$

The force law associated with this field intensity is verifiable for spherical masses. In this presentation however, it is assumed that it applies only to point masses: it will be shown below that the field whose source is an extended spherical mass is identical (outside the mass) to that generated by a point having the same mass located at its center. Finding the gravitational field generated by a given mass distribution is essentially a problem in space geometry. The fields can be expressed in closed form for a few of the simplest geometric configurations but generally (classical) fields are the solution to a linear partial differential equation and are expressed as infinite series. We will consider only a few simple cases here to demonstrate three definitive elementary methods for finding fields.

The first method is by direct calculation: the given mass is partitioned into differential size pieces; the contribution to the field at a point (x, y, z) from a typical piece is calculated; the contributions from these pieces are added together (integrated). Note however that the contribution from one such piece is a differential **vector** which does not point in the same direction as the other differential contributions: it is therefore necessary to resolve each contribution into components before adding them. For example, if Cartesian coördinates are used then the x-components of the contributions must be added together to find the net x-component of the field at the point in question and similarly for the y- and z-components.

The second method is related to the potential energy of the gravitational field. Remember that it is possible to define a potential energy V for the gravitational field (see *section* $1.3.4.3$.):

$$V = -G\frac{mm'}{r}$$

Since the gravitational field is the force divided by m', it follows that a potential function W can be defined analogously:

$$W = -G\frac{m}{r}.$$

It was possible to define V only because the work done by the gravitational field was independent of the path of motion. Furthermore, since the potential energy was in a sense the work integral already evaluated and the potential energy is the line integral of the force along the path of motion, it was suspected that the force must be some kind of derivative of the potential energy: upon investigation of this idea, it was found that the force in a given direction was the (negative of) the directional derivative of V in that direction

$$F_s = -\frac{\partial V}{\partial s},$$

where ds is a differential length in the s direction. It follows that the field in the direction of ds is

$$U_s = -\frac{\partial W}{\partial s}$$

This can be carried a little further. The force vector $\vec{F}$ in Cartesian coördinates is

$$\vec{F} = F_x\vec{i} + F_y\vec{j} + F_z\vec{k}$$

$$\vec{F} = -\left(\frac{\partial V}{\partial x}\vec{i} + \frac{\partial V}{\partial y}\vec{j} + \frac{\partial V}{\partial z}\vec{k}\right)$$

The quantity in parentheses is called the gradient of V. Hence the rule, 'the force is the negative of the gradient of the potential energy'.

Analogously,

$$\vec{U} = U_x\vec{i} + U_y\vec{j} + U_z\vec{k}$$

$$\vec{U} = -\left(\frac{\partial W}{\partial x}\vec{i} + \frac{\partial W}{\partial y}\vec{j} + \frac{\partial W}{\partial z}\vec{k}\right);$$

and the gravitational field is the negative of the gradient of the potential.

Notice that V and W are distinguished in that V is referred to as the potential energy and W is referred to as the potential.

This implies another general method for finding fields. The given mass is divided into differential pieces as before and the contribution of a typical piece to the potential at a point (x, y, z) is found; then the contributions are added to

447

find the total potential at that point. (Since the potential is a scaler, there are no components to deal with: adding the contributions to the potential is a scaler addition). The field is then the negative gradient of the potential (or if you prefer, the field in any given direction is the negative of the partial derivative in that direction).

The third method to be discussed depends on a theorem due to Gauss pertaining to the flux across a closed surface. There is a discussion of this in *section* $3.3.1.2$. This theorem will be discussed in detail later but briefly, the assertion is that the total flux of the gravitational (electrostatic) field across a closed surface is proportional to the field source, the mass (charge), enclosed by the surface.

The most important use of the theorem is its application to theoretical field theory. An example is its application to a differential region of space which results in the differential equation satisfied by the potential for any field in terms of the local mass density. This analysis will be presented later. In some cases where there is a good deal of symmetry, the theorem can be used in its integral form to find the gravitational field.

Symmetry arguments are used whenever possible to reduce the amount of work necessary to find a field. As an example, consider the field whose source is a spherical mass. By symmetry, it is known at the outset that the field is radially directed because anything else offends the spherical symmetry inherent to the configuration. Arguments such as this one often save a good deal of work in the process of finding a field.

The gravitational fields will be found for the following geometric configurations of mass distribution: a ring, a disk, a solid sphere, a hollow sphere, a solid cylinder, a hollow cylinder and a dipole (two spherical masses close together). In the cases of the ring and the disk, only the field along the axis normal to the plane of the figure and passing through its center will be found. The cylinders will be considered to be very long to avoid issues at its ends. In the cases of the spheres and the cylinders, the fields will be found both inside and outside the boundaries of the object. Each of the fields mentioned will be determined by whatever method can be applied with reasonable ease. It will be seen that some solutions easily obtained by one method are more difficult to obtain by another.

Before continuing, it should be noted that the electrostatic force of attraction (or repulsion) is of the same form as the gravitational. Hence, everything stated above concerning gravitational fields applies equally as well to electrostatic fields. Also the fields to be found below have analogous electrostatic fields of the same configuration. Note that while one seldom finds a mass distribution in empty space that is in the shape of a ring or a cylinder, a charge distribution of either one of these shapes is not uncommon. So even in the cases of unlikely mass distributions, the problems are carried through, not only because they might be applicable to an electrostatic situation, but also because the point here is to illustrate the methods for finding fields. These unlikely configurations serve that purpose well.

6.3.3.2 **Applications of the Direct Method**

Example 1: The Field along the Axis of a Ring

Consider a ring of mass **m** and radius **R** and suppose that the field of this ring is to be found as a function of the distance **s** along the axis as shown in **figure 6.8.**

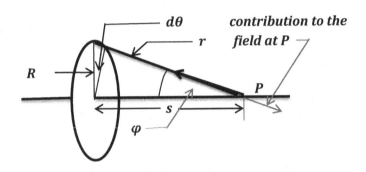

figure 6.7

Field Along the Axis of a Ring

Solution

The mass per unit length μ of the ring is

$$\mu = \frac{m}{2\pi R}.$$

The differential mass is (the mass per unit length μ, times the length $Rd\theta$)

$$dm = \mu R d\theta$$

$$dm = \frac{m}{2\pi} d\theta$$

Use the gravitation law to find the contribution of this typical piece:

$$dU \;=\; -G\frac{dm}{r^2} \;=\; -\frac{G\,m}{2\pi(R^2 + s^2)}\,d\theta$$

This differential contribution points in the direction of the arrow in the diagram. The component normal to the axis cancels with the one from the opposite side of the ring. The contributions above cannot be added together as they stand because they point in different directions. We take U_a, the component along the axis:

$$dU_a \;=\; \frac{-Gms}{2\pi(R^2 + s^2)^{3/2}}\,d\theta \;,$$

where the previous expression was multiplied by

$$\cos(\varphi) \;=\; \frac{s}{(R^2 + s^2)}$$

to single out the axial component.

The total component along the axis is found by adding all the differential axial components together (integrate). All the quantities on the right are constants except $d\theta$ and therefore can be taken out of the integral and we have

$$U_a \;=\; \frac{-Gms}{2\pi(R^2 + s^2)^{3/2}}\int_0^{2\pi} d\theta$$

$$U_a \;=\; \frac{-Gms}{(R^2 + s^2)^{3/2}}$$

Note that a typical point at an unspecified distance s along the axis was chosen. The result therefore pertains only to points along the axis. To obtain the field at off-axis points, one must, at the outset, pick an arbitrary (but unspecified) off-axis point P. In this case, the contributions from the differential pieces in the ring become a function of the angle at **P** subtended by the radius **R** of the ring and the integral becomes very difficult to evaluate. This is primarily a math problem and is avoided here.

Note also that as one moves very far from the source of the field (i.e. as **s** → ∞), the ring looks like a point source. It would therefore be expected that the field look like the field from a point source. This is in fact the case:

$$\left(\frac{-Gms}{(R^2 + s^2)^{3/2}}\right) \text{ is assyptotic to } \left(\frac{-Gm}{s^2}\right) \text{ as } s \to \infty,$$

which is the expected expression. (Hint: since **R** ≪ **s**, neglect the **R**.)

This result will now be used to develop the formula for the field of a solid sphere at a point outside the sphere. This will be accomplished in two steps:

1. *the field of a solid disk will be found by dividing the disk into differential concentric rings and adding the contribution of each of them using the formula found above;*
2. *the field of a solid sphere will then be found by partitioning the sphere into a stack of disks of differential thickness and adding the contributions from each of these using the formula obtained from step 1.*

It will be seen that the field of a sphere outside the mass **m** is identical to that of a point mass **m** concentrated at its center. Notice that only the formulas for the fields along the axis of the ring and the disk are used in this development. However, the symmetry of the final spherical distribution of mass implies that the result is valid for all points outside the sphere regardless of the direction.

The method for evaluating the integral in step **2** is not obvious. Even if you cannot follow the details through, know that it is more important to follow the method of breaking down the problem into simpler problems. Keep your attention in the right place; don't get swallowed up by the details of evaluating the integral. That part of the problem comes under the heading of *'methods of integration'* and is part of any course in elementary calculus.

Example 2: The Field along the Axis of a Disk

Find the field along the axis of a solid disk.

Solution

Let **m** be the mass of the disk and **R** its radius. *Figure 6.8* depicts a typical differential ring of radius **x** and thickness **dx**. **P** is a point along the axis a distance **s** from the origin which is taken at the center of the disk.

The mass **σ** per unit area of the disk is

$$\sigma \;=\; \frac{m}{\pi R^2}$$

The mass **dm** of the ring is (the mass per unit area **σ**, times the area **2 πxdx**)

$$dm \;=\; \sigma 2\pi x dx$$

$$dm \;=\; \frac{2mx}{R^2}\, dx$$

The differential contribution to the field at **P** from the ring is (see previous problem)

453

$$dU = -\frac{2Gmsxdx}{R^2(x^2 + s^2)^{3/2}}$$

$$U = \frac{-Gms}{R^2} \int_{x=0}^{x=R} \frac{d(x^2 + s^2)}{(x^2 + s^2)^{3/2}}$$

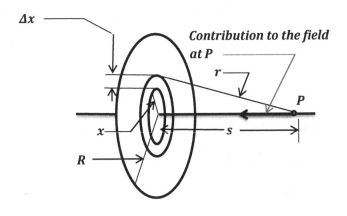

figure 6.8

Field Along the Axis of a Disk

Let $v = x^2 + s^2$. Then the integral to be evaluated is

$$U = \frac{-Gms}{R^2} \int_{x=0}^{x=R} v^{-3/2} \, dv$$

$$U = \frac{2Gms}{R^2} v^{-1/2}$$

$$U = \frac{2Gms}{R^2}\left(\frac{1}{(R^2 + s^2)^{1/2}} - \frac{1}{s}\right)$$

As a partial check, let $s \to \infty$: the above expression should approach that of a point source.

$$U = \frac{2Gm}{R^2}\left(\frac{1}{\left[1 + [R/s]^2\right]} - 1\right)$$

$$U \approx \frac{2Gm}{R^2}\left(1 - \frac{1}{2}\frac{R^2}{s^2} - 1\right)$$

$$U \approx -\frac{Gm}{s^2}$$

which is the expression for the field of a point source.

The following approximations were used in the above. For small x:

$$\sqrt{1 + x} \approx 1 + \frac{x}{2};$$

$$\frac{1}{1 + x} \approx 1 - x.$$

The problem now is to partition a sphere into a stack of disks as shown in *figure 6.9*. We will add the contribution of each disk using the last result.

Example 3:   The Field of a Solid Sphere outside the Sphere

Find the field of a solid sphere of radius T in the region outside the sphere.

Let T be the radius of the sphere, P the point along the axis where the field will be computed, x the distance from P to the center of the sphere, s the distance from P to a typical disk, R the radius of the typical disk, and θ the angle between the radius T and the x-axis as shown in *figure* 6.9.

The mass per unit volume λ of the sphere is

$$\lambda \quad = \quad \frac{3m}{4\pi T^3}.$$

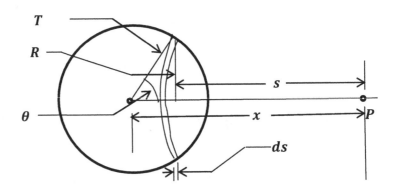

figure 6.9

Field of a Solid Sphere

Solution:

The mass of the disk is λ times the volume of the disk:

$$dm \quad = \quad \frac{3mR^2}{4T^3} ds$$

and, from the previous result, the contribution of this disk to the field at P is

$$dU = \frac{3Gm}{4T^3}\left(\frac{1}{R^2 + s^2} - \frac{1}{s}\right)sds$$

Notice that R is a variable. Therefore, it must be expressed in terms of s before integrating. From the diagram

$$R^2 = T^2 - (x - s)^2$$

To add the contributions from the disks, integrate from $s = x - T$ to $s = x + T$. Denote the two terms on the right by U_1 and U_2. Then the second term is

$$U_2 = -\frac{3Gm}{2T^3}\int_{x-T}^{x+T} ds$$

$$U_2 = -\frac{3Gm}{T^2}$$

The first term needs a little work. First eliminate R using the relation above:

$$U_1 = \frac{3Gm}{2T^3}\int_{s=x-T}^{s=x+T} \frac{sds}{\left(2xs - [x^2 - T^2]^{1/2}\right)}$$

$$U_1 = \frac{3Gm}{2T^3(2x)^2}\int_{s=x-T}^{s=x+T} \frac{(2xs)d(2xs)}{(2xs - [x^2 - T^2])^{1/2}}$$

The integral (call it K), is of the form

$$K = \int \frac{udu}{\sqrt{u - a}}$$

$$K = 2\int ud\sqrt{u - a}$$

$$K = 2u\sqrt{u - a} - 2\int \sqrt{u - a}\, du$$

$$K = 2u\sqrt{u - a} - \frac{4}{3}(u - a)\sqrt{u - a}$$

$$K \quad = \quad \frac{2}{3}(u + 2a)\sqrt{u - a}$$

Substituting back: $u = 2xs$ and $a = x^2 - T^2$ and simplifying

$$U_1 = \frac{Gm}{4T^3x^2}(xs + [x^2 - T^2])\sqrt{2xs - (x^2 - T^2)} \ \Big|_{\substack{s = x + T \\ s = x - T}}$$

$$U_1 = \frac{Gm}{2T^3x^2}[(2x - T)(x + T)^2 - (2x + T)(x - T)^2]$$

$$U_1 \quad = \quad \frac{Gm}{2T^3x^2}[6x^2T - 2T^3]$$

The total field at P is

$$U \quad = \quad U_1 + U_2$$

$$U \quad = \quad \frac{3Gm}{T^2} - \frac{Gm}{x^2} - \frac{3Gm}{T^2}$$

$$U \quad = \quad -\frac{Gm}{x^2}.$$

This is the field of the sphere of mass m and radius T at a distance x ($x > T$) from its center. It is the same as that of a point mass m at the center of the sphere.

This justifies the analysis of the two-body problem in which the planets were replaced by point masses. The above result shows that the motion of spherical planets with arbitrary radii move according to our result in *section 6.2.3*.

Example 4: The Field outside a Hollow Sphere

Find the field of a hollow sphere of mass m and radius R.

Discussion

In spite of the hairy integral in the previous analysis, things are working out pretty nicely here: it has been shown that the field from a spherical mass m is the same as that of a point mass m all concentrated at the center. At least that is the situation for points in space outside the sphere. What goes on inside the sphere will be discussed later.

We're going to pull a fast one now to find the field outside a hollow sphere: our trick will be to start with the field of a solid sphere of mass M and radius R and subtract the field of a smaller concentric sphere of mass M' and radius R'. This is equivalent to hollowing out the original sphere. Let m be the mass of the final spherical shell.

Solution

Since the centers of the two masses coïncide, the field at a distance r from their common center is the field of the original sphere minus the field of the hollowed out portion:

$$U = -G\frac{M}{r^2} + G\frac{M'}{r^2}$$

$$U = -\frac{G}{r^2}(M - M')$$

This implies that the field from the remaining shell is

$$U = -G\frac{\Delta M}{r^2},$$

$$U = -G\frac{m}{r^2}$$

where $\Delta M = m$ is the mass of the shell. Hence outside the sphere, the field is still the same as that of a point mass.

Example 5: The Field inside of a Hollow Sphere

The field inside the shell is a different matter. A simple geometric argument shows that the field inside the spherical shell is zero. *Figure 6.10* depicts a hollow sphere and an arbitrary point P inside.

Let $d\Omega$ be a differential solid angle with its vertex at P. and let dA and dA' be the intersections of the angle with the surface at respective distances r and r' from P. The contributions of the two areas are in opposite directions. If σ is the mass density on the surface, then the contribution to the field at P is

$$dU = -G\frac{\sigma dA}{r^2} + G\frac{\sigma dA'}{r'^2}$$

Each differential area on the surface is proportional to the square of its distance from P. If k is the proportionality constant, then

$$dU = -G\frac{\sigma kr^2}{r^2} + G\frac{\sigma kr'^2}{r'^2}$$

$$dU = 0$$

Since P and $d\Omega$ were arbitrary, it follows that the field everywhere inside the hollow sphere is zero.

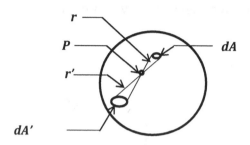

figure 6.10

Field inside a Hollow Sphere

Usually, the use of potentials simplifies problems in field theory, but this particular problem is the exception to that rule. This problem is more difficult to solve using potentials.

Example 6: The Field of a Mass Dipole

Two equal masses each of mass m are placed symmetrically around the origin, one at $x = a$ and the other at $x = -a$. Find the field from this configuration. (This configuration is referred to as a mass dipole.)

Discussion

The field from the dipole has cylindrical symmetry around the x-axis. It is therefore sufficient to consider only points in the xy-plane. The configuration is shown in *figure* 6.11. Let P be an arbitrary point (x, y) in the plane. Because the field is dependent on distances, the expressions in Cartesian coördinates are fraught with

461

Pythagorean square roots. These look hairy but for the most part can be written down by inspection.

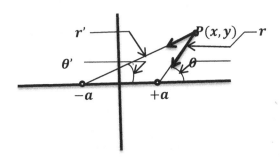

figure 6.11

Field of a Mass Dipole

From the diagram, it follows that

$$U_x = -Gm\left(\frac{\cos(\theta)}{r^2} + \frac{\cos(\theta')}{r'^2}\right).$$

Write the cosines as

$$\cos(\theta) = \frac{x-a}{r} \quad \text{and} \quad \cos(\theta') = \frac{x+a}{r'}.$$

Then the **x**-component of the field is

$$U_x = -Gm\left(\frac{x-a}{[(x-a)^2+y^2]^{3/2}} + \frac{x+a}{[(x+a)^2+y^2]^{3/2}}\right)$$

Similarly, the y-component is

$$U_y = -Gm \left(\frac{y}{[(x-a)^2 + y^2]^{3/2}} + \frac{y}{[(x+a)^2 + y^2]^{3/2}} \right)$$

In the next section, the field potential for this configuration will be obtained and these expressions will be its directional derivatives in the x- and y-directions respectively. The difference of the two methods is well illustrated by this problem

Note: 'Dipole field' often refers to an approximation of these expressions when a<<x, y. This issue will not be carried through beyond this point here because it is just a mathematical exercise.

6.3.3.3   Applications of the Potential Method

Recall that the potential energy was the integral of the force along some path and could be defined only when the integral over a position change from point **A** to a point **B** was independent of what path was chosen. This means essentially that the potential is some sort of an integrated version of the force and that the force could therefore be recovered from the potential by some sort of a derivative. The advantageous thing about this is that the potential is a scaler quantity and hence, easier to deal with than the vector force field. This was carried further and it was found that the force in any direction could be recovered by taking the derivative of the potential in that direction.

When attention is focused on the gravitational field rather than the gravitational force, the only difference is the absence of one of the masses in the basic formula. While this leads to considerable philosophical difficulties in terms of the existence or non-existence of fields, it is laughably trivial in terms of mathematics. Consequently, potentials are definable for all those fields that are associated with conservative forces and potentials are in the same relation to fields as potential energies are to force fields.

Potential theory starts with the potential from a point source:

$$W \quad = \quad -G\frac{m}{r},$$

where W is the potential function whose source is the mass m a distance r away. To find the potential function for a mass distribution, one divides the distribution into infinitesmal chunks and adds the contribution of each chunk. This is usually easier than the direct method of the previous section because the potential is a scaler function and is found by a scaler addition (integral): after it has been found, the actual field can be determined by taking directional derivatives which is a comparatively straightforward process.

Each of the fields in the previous section will now be found again using potentials. A comparison should be made to acquire a 'feel' for each of the two methods and the difference between them.

Example 1': The Potential along the Axis of a Ring

Find the potential along the axis of a ring of mass m and radius R. Take the derivative in the direction of the axis to determine the field in that direction.

Discussion

We are going to add the scaler contributions to the potential of each of the differential pieces of mass around the ring. Since all the mass is at the same distance from any given point P on the axis, this is particularly easy to do. It can be done by inspection.

Solution

All the mass is at a distance r from the point P on the axis. Since

$$r = \sqrt{s^2 + R^2},$$

we have, by inspection

$$W = -G\frac{m}{(s^2 + R^2)^{1/2}}$$

In the event that you missed the fact that this could be done by inspection, the whole process is as follows. The mass per unit length μ of the ring is

$$\mu = \frac{m}{2\pi R}$$

The mass subtended by a differential angle $d\theta$ is the mass per unit length times the length $Rd\theta$:

$$dm = \frac{m}{2\pi R}Rd\theta$$

$$dm = \frac{m}{2\pi}d\theta,$$

and its contribution to the potential is

$$dW = -G\frac{dm}{r}$$

$$W = -G\frac{m}{2\pi(s^2 + R^2)^{1/2}}\int_0^{2\pi} d\theta$$

$$W = -G\frac{m}{(s^2 + R^2)^{1/2}},$$

which agrees with the answer obtained by inspection.

In the previous section, it was stipulated that when **s** is very large, the field look like that of a point mass. The situation with the potential is similar: the potential should look like that of a point mass.

Exercise

Show that for large **s**, the potential looks like

$$W \quad = \quad -G\frac{m}{s}$$

To find the field in the direction of the axis, take the derivative with respect to s. this is justified because the problem was set up for arbitrary points on the axis. This restriction to points along the axis however, precludes any statements concerning off-axis points. Therefore it is not legitimate to take derivatives in any other direction.

For the field in the direction of the axis, we have

$$U_s \quad = \quad -\frac{\partial W}{\partial s}$$

$$U_s \quad = \quad -\frac{Gms}{(s^2 + R^2)^{3/2}}$$

which agrees with the answer obtained in the previous section by the direct method.

Example 2′: The Potential along the Axis of a Disk

Find the field of a solid disk at points along the axis by dividing the disk into concentric rings and using the above result as the contribution of each of them. Find the field along the axis by taking the derivative with respect to **s**.

466

$$W \;=\; -G\frac{m}{s}$$

which is the potential for a point source. (Note: these asymptotic values are just a partial check.)

The field in the direction of the axis is the partial derivative of this expression with respect to s. Start with

$$W \;=\; \frac{-2Gm}{R^2}\left([s^2 + R^2]^{1/2} \;-\; s\right)$$

$$U_s \;=\; -\frac{\partial W}{\partial s}$$

$$U_s \;=\; \frac{2Gm}{R^2}\left(\frac{s}{[s^2 + R^2]^{1/2}} \;-\; 1\right)$$

$$U_s \;=\; \frac{2Gms}{R^2}\left(\frac{1}{[s^2 + R^2]^{1/2}} \;-\; \frac{1}{s}\right)$$

This result agrees with that of the previous section.

Example 3′: The Potential of a Sphere outside the Sphere

Find the potential for a sphere of radius T by stacking disks together to make a sphere. Take the derivative in the radial direction to find the field.

Discussion

Notice as you go along that this approach works out considerably easier than the direct method of the previous section.

Solution

The sphere is divided as before and the same parameters will be used here. The mass per unit volume λ of the sphere is

$$\lambda \quad = \quad \frac{3m}{4\pi T^3}$$

The mass of one of the typical disks is λ times the volume of the disk:

$$dm \quad = \quad \frac{3m}{4\pi T^3}\pi R^2 ds$$

$$dm \quad = \quad \frac{3mR^2}{4T^3}ds$$

Use this expression for the mass in the previously found formula for the potential W of a disk:

$$dW \quad = \quad -\frac{3Gm}{2T^3}\left([s^2 + R^2]^{1/2} - s\right)ds$$

The radius R of the disk however, is not constant during the integration process and must be eliminated before evaluating the integral:

$$R^2 \quad = \quad T^2 - (x-s)^2$$

Use this relation in the expression for dW and

$$W \quad -\frac{3Gm}{2T^3}\left(\int_{s=x-T}^{s=x+T} (2xs - [x^2 - T^2])^{1/2}ds - \frac{s^2}{2}\Big|_{s=x-T}^{s=x+T}\right)$$

$$W = -\frac{3Gm}{2T^3}\left(\frac{[2xs - x^2 + T^2]^{3/2}}{3x}\Big|_{s=x-T}^{s=x+T}\right)$$

$$-\frac{3Gm}{2T^3}\left(\frac{(x+T)^2}{2} - \frac{(x-T)^2}{2}\right)$$

$$W = -\frac{3Gm}{2T^3}\left(\frac{(x+T)^3 - (x-T)^3}{3x} - 2xT\right)$$

$$W = -\frac{3Gm}{2T^3}\left(\frac{6x^2T + 2T^3}{3x} - 2xT\right)$$

$$W = -G\frac{m}{x}$$

This last expression is the same as the potential for a point mass m at the center of the sphere. These calculations constitute the second proof that the field outside a spherical mass is identical to that produced by a point of equal mass at its center. The field inside a (solid) sphere is discussed later.

What we have is the potential at a distance x from the center of the sphere $(x > T)$. To find the field, take the derivative:

$$U_x = -\frac{\partial W}{\partial x}$$

$$U_x = -G\frac{m}{x^2},$$

which is the same as the field from a point source.

Example 5′ The Potential of a Mass Dipole

Find the potential for a mass dipole consisting of two masses each of mass m placed along the x-axis, one at $x = -a$ and the other at $x = a$. Find the x- and y-

components of the field from the potential by taking derivatives.

Solution

Refer to *figure 6.11*. Using Pythagoras' theorem, the potential can be written by inspection.

$$W = -Gm\left(\frac{1}{([x-a]^2 + y^2)^{1/2}} + \frac{1}{([x+a]^2 + y^2)^{1/2}}\right)$$

Exercise

The rest of this problem is yours: take the partial derivatives with respect to x and y to find the x- and y-components of the field, respectively. Compare your results with those obtained in the previous section.

6.3.4 The Law of the Conservation of Something or Other

6.3.4.1 Flux and Gauss' Law

In this section, we will develop the concept of flux which is most easily understood in terms of the flow of an incompressible fluid across a boundary. Water will serve very well as an object of consideration in this context.

Imagine a tank of water equipped with thin hoses which can inject more into it or drain some out of it. Also imagine that the ends of these hoses are at various points in the middle of the body of water that is already there.

Assume for the moment that the water flows into the tank via one hose at a constant rate of w cubic inches per second. The point where the hose ends is referred to as a *source of strength w,* because in terms of the flow of water which is the point of focus here, that is the point where water is added to the flow. Also assume that water leaves the flow via another hose at the rate of **k** cubic inches per second. The point where it leaves, i.e. the end of the draining hose is referred to as a *sink of strength k* because that is the point wher364e water is subtracted from the flow.

Now, imagine a closed surface of (almost) any shape (no Klein bottles or torus', please). This imagined surface should be a simple closed one with an inside and an outside and the properties:

1. *that you cannot get from one to the other without crossing the surface;*
2. *that the volume enclosed by it is finite.*

The issue to be addressed here is to account for the flow in and out of the region enclosed.

Let *S* represent the (fixed) surface, and *R*, the region interior to it. Suppose first that there are no sources or sinks enclosed by *S*. Then the water flowing into *R* across the surface *S* must equal the water flowing out of *R* across the surface *S*. This is because water is pretty incompressible stuff: it cannot rarefy or pile up so the amount of water inside *R* is constant. What we have here is the law of the conservation of water: it is neither created nor destroyed during its motion; what goes in must come out; or however you want to say it. The point is, don't lose the simplicity of this idea as we complicate matters a little.

Suppose now that there is a source of strength *w* enclosed by *S*. Then the net flow out of the region *R* is *w*. If, in addition, there is a sink of strength *k*, the net flow out of *R* has to be $(w - k)$. It becomes clear after considering several situations that it is necessary to know something about the net flow out of R. This is the flow *out* across the surface minus the flow *in* across the surface.

The concept of flow across a surface or boundary is called *flux.* When speaking about population movements into and out of a region, the words *'in-flux'* and *'out-flux'* are used. These terms are sometimes used in this context but often the single word *'flux'* covers both. *Out-flux* is taken as a positive quantity and *influx* as a negative out-flux. The net flux is the out-flux minus the in-flux. As seen above the flux across a closed surface is related in a simple way to the sources and sinks enclosed by the surface.

The missing element in all this is the mathematical description of flux that relates it to the flow parameters. To remedy this, consider a differential piece dS of the surface S as shown in *figure* **6.12**. Let $\vec{v}$ be the velocity of the flow at this location. Consider how much water is going to cross the differential piece of surface in the next second. Well, in terms of getting across the surface, the component of the velocity parallel to the surface is useless; only the normal component of velocity is pertinent. We need to know how much will cross the differential area per second.

Now, in grammar school, someone undoubtedly wrote on the blackboard

$$DISTANCE \quad = \quad RATE \quad \times \quad TIME$$

That's all you need. The water at the differential surface area will travel up a distance of v_n (the normal component of velocity) in one second. The amount of water that flowed across the surface in one second therefore, is the water in the rectangular parallelopiped formed by dS and v_n: its volume is the differential piece of surface area times the normal component of the velocity v_n. That product is the differential flux, df, across the differential piece of surface area:

$$df \;=\; v_n dS.$$

We can condense the notation somewhat if we define a vector which has to do with the geometry of the surface S: define the vector $\vec{dS}$ whose magnitude is the area dS and whose direction is normal to the surface pointing outward from the region R enclosed. Then the flux df, across the differential area can be written

$$df \;=\; \vec{v} * \vec{dS}.$$

The dot product neatly singles out the component of velocity in the direction of $\vec{dS}$ which was defined normal to the surface.

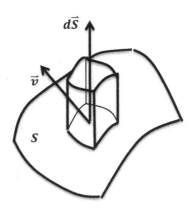

figure 6.12

The Flux Across a Differential Piece of Surface

So far, so good. To find the amount of water flowing out from the region R enclosed by the surface, these differential contributions must be added over the whole surface. By this time you must have gathered that the process of adding differential contributions of a quantity to obtain the total quantity means that you integrate. It is no different in this case: the total flow out of the enclosed region is the integral (sum) of the differential contributions described above. While you may not know how to evaluate these integrals (pending the acquisition of more prowess in calculus) the symbols (presented below) used to describe the operation of finding the flux are pretty descriptive of the process and should be understood.

The symbols can be described as follows. The differential area is two dimensional and therefore, two integrations are needed to evaluate the flux over the whole surface S. Hence, the symbol for the flux has two integral signs. They are subscripted with an S to indicate that the integral is to be carried out over the surface S. A circle is drawn across the middle of the integral signs to indicate that the surface S is closed. Put all these elements together and the symbol adopted to represent the flux of water across the surface is

$$f \;=\; \int df \;=\; \oiint_S \vec{v} * d\vec{S}$$

Now that's a pretty spiffy notation. I wish I had thought of it myself. But don't be mystified or put off by its appearance. Once again, although you may not know how to carry out the operations implied by the expression on the right, the symbol represents a very simple idea; it represents the total volume of water per second that flows out of the region by crossing the enclosing surface S. You should be able to translate the symbols into English even if you cannot perform the mathematical operations they indicate.

This topic is now going to be developed further: the other side of the equation that pertains to the sources and sinks is in need of a description.

First consider the case when there are no sources or sinks. Then, because the water is incompressible, the volume of water inside the region R is constant which implies that there is no net out-flux (or in-flux). The equation that applies is

$$f = 0$$

$$\oiint_S \vec{v} * d\vec{S} = 0.$$

This does **not** mean that $\vec{v} = 0$: this equation can be satisfied because at some locations on the surface, water can be flowing into R making a negative contribution to the flux; in other locations, the water can be flowing out of R making a positive contribution to the flux. The equation **does** mean that these contributions must add to zero when they are all added together (integrated over the surface S). There are many flow configurations inside the tank that can satisfy this condition.

Next, suppose there are a number n of sources of strengths $w_1, w_2, \ldots, w_n$ and a number m of sinks of strength $k_1, k_2, \ldots, k_m$. Then the total amount of water w injected into the tank per second is

$$w = w_1 + w_2 + \cdots + w_n - (k_1 + k_2 + \cdots + k_m)$$

$$w = \sum_{i=1}^{n} w_i - \sum_{i=1}^{m} k_i.$$

In this case, the equation that applies is

$$f = w$$

$$\oint_S \vec{v} * d\vec{S} = \sum_{i=1}^{n} w_i - \sum_{i=1}^{m} k_i$$

Keep in mind that the last equation says nothing more complicated than the net amount of water per second flowing out of **R** equals the total amount of water per second injected into **R**.

There is another case to consider. If the sources and sinks are not positioned at isolated points but rather distributed continuously throughout **R**, then an expression representing the water thus injected is required.[25] To this end consider a differential volume, **dv** in **R**:

$$dv = dx\,dy\,dz$$

If the injection rate of water per unit volume of **R**, at a point (x, y, z) is designated $\rho(x, y, z)$, then the (differential amount of) water per second injected into the differential volume is

$$dw = \rho(x, y, z)dxdydz.$$

The total water per second **w** injected is the (three-dimensional) integral over **R** of all these contributions. It is written

---

[25] If you have trouble visualizing water injected over some continuous region, imagine a river flowing through a valley surrounded by hills. After a rainstorm, water drains into the river along its banks. If x is distance along one bank and $\mu(x)$ is the number of cubic inches of water draining into the river per inch of bank length at x, then the differential amount of water injected in a differential length dx is $dw = \mu(x)dx$ and the total injection is $w = \int_{length\ of\ bank} \mu(x)dx$. The issue addressed here is the same except the continuous injection is throughout a volume in the region R instead of along the one-dimensional length of river bank.

$$w \;=\; \iiint_R \rho(x,y,z)dxdydz$$

The three integral signs indicate that the integral is over a three-dimensional volume. The subscript R indicates that the volume of integration is the region R. The equation in this case is

$$f \;=\; w$$

$$\oiint_S \vec{v} * d\vec{S} \;=\; \iiint_R \rho(x,y,z)dxdydz$$

(Usually, in physics, if a closed surface integral is equal to a volume integral, the volume is the one enclosed by the surface.) Finally, note that the flux equation could have been written

$$\oiint \vec{v} * d\vec{S} = \sum_{i=1}^{n} w_i - \sum_{i=1}^{m} k_i + \iiint_R \rho(x,y,z)dxdydz$$

if there were point sources and sinks at some locations and continuous injections at others. (This last equation even covers the case where both injection types occur at the same point.)

The various cases above automatically take care of themselves: if there are no point sources and sinks, the $w_i's$ and the $k_i's$ are zero; in those regions where there is no injection spread out over a continuous region, the function $\rho(x,y,z)$ is zero. A little thought reveals that the above equation covers all possible configurations of sources and sinks correctly.

We will consider one last aspect of this subject. For this one, we leave the comfort of 'incompressibility'. Consider a gas in the container. For a gas flux, sources and sinks are no longer measured in terms of cubic inches per second but rather in mass per second. It sometimes occurs that what goes in **doesn't** come out. This can occur because gas is compressible: it can 'pile up' or rarefy in the region **R**. To describe this mathematically, you need the gas density in **R**. If $\sigma(x, y, z, t)$ represents the mass density at the point (x, y, z) at time t, then the total mass m of gas in **R** at some time t is the volume integral of $\sigma(x, y, z, t)$:

$$m(t) = \iiint_R \sigma(x, y, z, t) dx dy dz$$

What is inside the integral looks incomprehensible but it is simply the mass per unit volume $\sigma(x, y, z, t)$ at time t, times the volume $dx dy dz$.

If the flux (the flow out of **R**) exceeds the injection rate then the mass of the gas in **R** is decreasing at that many mass units per second. The excess flux out of **R** is the mass per second leaving **R**. If f_e represents the excess flux, then

$$f_e = -\frac{dm}{dt}$$

The full flux equation then becomes

$$\oiint_S \vec{v} * d\vec{S} = \sum_{i=1}^{n} w_i - \sum_{i=1}^{m} k_i$$
$$+ \iiint_R \left[\rho(x, y, z) \right.$$
$$\left. - \frac{\partial}{\partial t} \sigma(x, y, z, t) \right] dx dy dz.$$

In the compressible case, the mass density at a point (x, y, z) changes with time: if it didn't, the mass in **R** would remain constant. The partial derivative is used under the integral sign to indicate that before the volume integration the integrand refers to a typical fixed point (x, y, z) so that x, y and z are constant during this time-differentiation.

The volume integral then has the effect of adding the mass density changes throughout the entire volume.

Expressions like the last one have a certain foreboding to those who are not used to them. They seem to the novice to be completely incomprehensible and lead one to believe that they will always remain that way. ***This is not so.*** The previous development was carried out to illustrate that such expressions can be understood in terms of their translation into language even without the ability to perform the details of the mathematical operations indicated. To reiterate, the left side of the equation represents the net flow-rate across *S* (out of the region *R*); the two summations on the right side of the equation and the first term in the integral represent the net injection rate of the water. For incompressible fluids, these terms say that what flows out/sec is the same as what flows in/sec. The last term (with the time-derivative) is for compressible fluids: it accounts for the net decrease/sec of stuff in *R* when the outflow rate is more than the inflow rate. Don't let the scary looking symbols make you lose sight of these simple ideas.

This development was carried much further than was necessary for our purposes. Only the first two terms were needed: the flux out of *R* and the sources in *R*. The rest serves to illustrate how verbal statements are made into mathematical ones. The hope is that you will acquire some prowess in translating such expressions back into English and perhaps even learn a little about formulating them.

6.3.4.2 Back to the Business At Hand

Consider a special case: imagine water injected into the center of a sphere and moving away from the injection point in a uniform, spherically symmetric manner.[26] Let the strength of the source be **w**. For the flux equation we have,

$$f = w$$

$$\oiint_S \vec{v} * d\vec{S} = w$$

You thought that there was no way you could evaluate a flux integral. Here's one situation where you can. From symmetry we know that the velocity is normal to any spherical surface centered at the source point. Also, $d\vec{S}$ on the surface points radially outward from the surface so the angle between the velocity and the differential surface vector is zero. Since we are free to choose any surface we want to find the flux, pick a sphere of (any) radius r centered at the source **w** because that makes things simple.

The dot product is just the product of the magnitudes:

$$\oiint_S |\vec{v}||d\vec{S}| = w$$

The magnitude of the velocity is constant over the sphere's surface (by symmetry), so it can be taken out of the integral:

$$|\vec{v}| \oiint_S |d\vec{S}| = w$$

What is left under the integral sign is the differential surface area. Its integral is just the surface area of the sphere chosen:

---

[26] This 'special case' of uniform flow out from the source point is preparing for the analysis of the flux of the gravitational field: even though velocity fields do not normally behave in this organized a fashion, the gravitational field of a point mass does.

$$|\vec{v}|\, 4\pi r^2 \quad = \quad w$$

Solve for the magnitude of the velocity:

$$|\vec{v}| \quad = \quad \frac{w}{4\pi r^2}$$

The magnitude of the velocity falls off as the square of the distance from the source. That makes sense. The source *w* is constant so the flux must be constant. But the flux is essentially the magnitude of the velocity times the surface area. As you move away from the source, the surface area increases as the square of the distance; it follows that the velocity decreases in magnitude as the square of the distance so their product remains constant. The point is that in space,

> *an inverse square dependence with distance from some central source point is the natural 'thinning out' process that results from the geometry of three dimensions.*

Newton had the right idea with his corpuscular theory of gravitation. Suppose corpuscles are converging on a spherical mass from outer space. Then the inverse square dependence of the gravitational attraction law is explained by the natural increased density of the corpuscles as they converge on the body; there is no necessity to increase or decrease the number of corpuscles en route or to change their individual impact. It is all taken care of by the geometry of their convergence. This in fact was the original appeal of the theory. (See *section* **3.3.1.2** for a discussion of this.)

The important result is that there is no escaping the fact that in three dimensions, inverse square dependence is a geometric phenomenon. Furthermore, it is the only type of dependence in three dimensions that can be purely geometric.

We spoke here of the law of the conservation of water; Newton conjectured the conservation of corpuscles. In the presence of an inverse square law, it is difficult to avoid the feeling that something is conserved because given that the material witch is flowing is conserved, i.e. that it is neither created nor destroyed en route, the geometry of three dimensions accounts for this particular dependence.

We have incidentally derived Gauss' law in the course of the above discussion. It is basically the law of incompressible stuff: the sum of the sources in any region is equal to the flux across the boundary of the region.

We didn't need water and flows involving a velocity vector to define a flux. Indeed newton didn't need corpuscles. All that was necessary was the vector field and a surface. Since we supply the surface, all that is ultimately needed is a vector field.

Essentially,

> *if a vector field from a point source has an inverse square dependence and the field satisfies the superposition principle then for any collection of sources and sinks, discrete and/or continuous, the resulting field satisfies Gauss' law.*

The only thing that might need explanation is the 'superposition principle.' A vector field is said to satisfy the superposition principle if the field of a collection of sources is the vector sum of the fields from the same sources taken individually. This principle was tacitly assumed to hold for gravitational fields throughout this chapter every time we added field contributions vectorially to obtain a total field.

It is important to note that Gauss' law is a geometric law and applies only if the vector field from a point source falls off according to the natural 'thinning out' process implied by the geometry. For three dimensions this means an *inverse square dependence.* The tidy relation between sources enclosed by, and flux across an arbitrary surface does not hold for any other dependence. For example, in two dimensions (maybe water flowing onto a flat surface from a point where it bubbles up) the natural geometric 'thinning out' is inversely proportional to the distance from the source: hence, an inverse first power law would allow for an analogue of Gauss' law. It would be stated, 'the net flux of water across an arbitrary curve enclosing the source would equal the amount of water bubbling up'.

To determine the constant that applies to the gravitational equation, consider a sphere of radius **R** enclosing a spherical source mass **m** and apply the flux equation:

$$\oiint_S \vec{U} * d\vec{S} \quad = \quad W$$

Like before the angle in the dot product is zero and the symmetry makes $|\vec{U}|$ constant over the surface:

$$\oiint_S \frac{Gm}{R^2} |d\vec{S}| \quad = \quad W$$

$$\frac{Gm}{R^2} \oiint_S |d\vec{S}| \quad = \quad W$$

$$\frac{4\pi R^2 Gm}{R^2} \quad = \quad W.$$

Therefore

$$W \ = \ 4\pi Gm,$$

and the Gaussian form of the gravitational law is

$$\oiint_S \vec{U} * d\vec{S} \ = \ 4\pi Gm$$

The factor of **4π** in this expression comes from the geometry of spheres and appears in Gauss' law or not depending on the definition of **G**. **(G** is sometimes defined so as to incorporate it.)

It should be remembered that although simple sphere geometry was used extensively in the above analyses, there was no loss of generality. As implied at the beginning, the relation of the flux to the sources enclosed holds for arbitrary surfaces.

6.3.5 Applications of Gauss' Law

The integral form of Gauss' law is not well suited to find particular fields, but it is highly valuable as a theoretical tool as will be seen in *section* **6.4**. There are a few cases however where it can be used in the context of determining fields. These are the cases for which symmetry allows you to choose a flux surface over which you know from the symmetry, that the field **U** is constant. Then it can be taken out of the integral and what is left inside is the differential surface area which integrates to the area of the surface so chosen.

In all cases considered here, the field **U** will be either normal to the surface or parallel to it: in either case it will not be necessary to consider the angle in the dot product: the cosine will be either zero, one or minus one.

The Field of a Point Mass

We apply the law first to a point mass. Remember that it was the basic inverse square law of Newton that told us that Gauss' law is applicable. Hence, we are not verifying the gravitational law for point masses: that was the experimental basis for all this. We are applying it to check the consistency of our form of Gauss' law with the experimentally determined one.

Consider a sphere of radius r whose center is at the point mass m. Gauss' law then states

$$\oiint_S \vec{U} * d\vec{S} \;=\; 4\pi G m.$$

In this case, $\vec{U} * d\vec{S} = -|\vec{U}||d\vec{S}|$. By symmetry, $|\vec{U}|$ is constant over the surface and can therefore be taken out of the integral:

$$-U_r \oiint_S |d\vec{S}| \;=\; 4\pi G m$$

$$-U_r 4\pi r^2 \;=\; 4\pi G m$$

$$\Rightarrow \quad U_r \;=\; -G\frac{m}{r^2},$$

which is the correct form. The minus sign enters the equation because the differential area vector was defined positive pointing outward and the field is attractive and hence points inward.

The Field of a Spherical Mass

Neither the ring nor the disk has symmetries which allow for an analysis using the integral form of Gauss' law. However, the sphere of mass m and radius R does. Recall that it was necessary to go through the ring and disk problems to show that outside a spherical mass m, the gravitational field is the same as that of a point mass m at its center. This result follows immediately from Gauss' law. We have

$$\oiint_S \vec{U} * d\vec{S} = 4\pi Gm.$$

As long as the spherical surface chosen to apply the law has radius r greater than R, the above equation is solved for the field in precisely the same way as for the point mass.

The Field inside the Sphere

Can Gauss' law be applied to the inside of a solid spherical mass? The answer is 'yes'. Pick a concentric spherical surface of radius $r < R$. Symmetry still allows that U_r be taken out of the integral but the mass that occurs on the right side of the equation is the mass **enclosed by the surface.** The equation becomes

$$-U_r 4\pi r^2 = 4\pi G \frac{r^3}{R^3} m$$

$$\Rightarrow \quad U_r = -G \frac{mr}{R^3}$$

This is a new result. It says that the field is zero when $r = 0$ and increases linearly from the center until you get to the surface $(0 \leq r \leq R)$, and then drops off inversely as the square of the radius r for $(R \leq r \leq \infty)$. The formulas for inside and outside the mass match at the boundary. This result is not easily obtained by a direct application of the other methods discussed.

The Field inside a Hollow Sphere

The field inside a hollow sphere is easily found using Gauss' law. Simply note that a concentric sphere inside the hollow one satisfies the symmetry condition and does not enclose any mass: hence, the right side of the equation is zero and the field is also.

The Field outside a Cylinder

As a final example, we will find the (approximate) field whose source is a long solid cylinder using Gauss' law. Let τ be the mass per unit length of the cylinder and R, its radius. Use will be made of coaxial cylinders of radius r with $r \geq R$ for outside the cylinder and $r \leq R$ for inside the cylinder. The surfaces we use will be of length L and we will stay away from the ends of the cylinder so that the flux across the ends of our imagined surfaces will be negligible. (Refer to *figure* **6. 13**.)

The mass enclosed is τL. By symmetry, the field is (mostly) radial so the flux across the ends can be neglected (the field is parallel to the surface at the ends). The radial component of the field is anti-parallel to the differential surface area. It follows that

$$\oint\!\!\!\!\oint_{side} \vec{U} * d\vec{S} \; + \; \oint\!\!\!\!\oint_{ends} \vec{U} * d\vec{S} \;\; = \;\; 4\pi G\tau L$$

As long as we stay away from the ends of the given cylinder, the second integral is negligible. By symmetry, the radial component of the field is constant over the side. Therefore,

$$-U_r 2\pi r L \quad = \quad 4\pi G \tau L$$

$$\Longrightarrow \quad U_r \quad = \quad \frac{-2G\tau}{r}$$

Hence, the field falls off as a $1/r$ function. That makes sense because the field spreads out in only one direction.

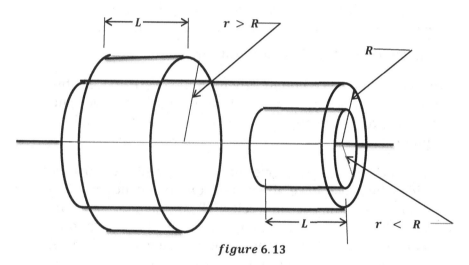

figure 6.13

The Field of a Solid Cylinder

The Field inside a Solid Cylinder:

To find the field inside the cylinder, use as your surface a coaxial cylinder with $r \leq R$. Once again, neglect the flux across the ends. The only difference in the analysis is that the mass **m'** enclosed is

$$m' = \frac{r^2}{R^2}\, \tau L$$

The result then, is

$$U_r 2\pi r L = -4\pi G\tau L \frac{r^2}{R^2}$$

$$\Rightarrow \quad U_r = \frac{-2G\tau r}{R^2}$$

As with the sphere, the field is zero at the axis and increases linearly with the radius r until you reach the surface, then it drops off inversely proportional to r.

The Field inside a Hollow Cylinder

If the cylinder is hollow, then in the case $r \leq R$, the Gaussian surface does not enclose any mass (charge) and the field turns out to be zero as in the case of the hollow sphere.

The examples given above apply mostly to electrostatic fields which also have an inverse square dependence: one seldom encounters in celestial mechanics, a long cylindrical mass large enough to make its gravitational field an object of interest but one often encounters wires, pipes and shafts which carry an electric charge.

Be aware that Gauss' law always holds for static inverse square fields. The difficulty in applying it in integral form arises from the fact that it is difficult in general to choose a surface over which the field is known to be constant. That is why symmetry was necessary to arrive at the results obtained above.

On the other hand, Gauss' law determines the differential equations for all inverse square fields. These can always be solved in one way or another. Infinite series solutions and numerical approximation solutions are often used because field problems are intricate space geometry problems and do not give up their secrets easily. Indirectly then, in the form of these differential equations, Gauss' law can be said to have universal applicability to inverse square fields.

6.4   THE DIFFERENTIAL EQUATION FOR STATIC INVERSE SQUARE FIELDS

6.4.1 The Derivation of the Equation

The differential equation that all inverse square fields must satisfy is the starting point for all field theory problems which cannot be solved by the elementary methods described above. The basic equation for static fields will now be derived. It is not the purpose here to solve this equation but merely to demonstrate that such an equation exists, to illustrate that it is determined by Gauss' law, that it is connected to the basic ideas presented here and to mention some of its general properties.

The equation in question is obtained by applying Gauss' law to an arbitrary differential region of space. What we will derive here is the equation expressed in Cartesian coördinates. *Figure 6.14* is a diagram of the differential region. It is just a box with differential dimensions dx, dy and dz. The procedure is to find the flux of the gravitational field across its surface and to relate that via Gauss' law to the mass enclosed.

Consider the faces A and B of the box as shown. $A's$ position in the horizontal direction is x; $B's$ position in the horizontal direction is $(x + dx)$. Both faces are rectangles with dimensions dy and dz and hence, have area $dS = dydz$. Since the y- and z-components of the field are parallel to these two faces, they do not contribute to the flux. Therefore, consider only the x-component U_x of the field.

In calculating the flux, notice that only the component of the field which is normal to the faces A and B enters into the calculation, so the cosine of the angle in the flux integral is $+1$ at face B, and -1 at face A. This is because the differential surface area is defined normal to the surface pointing outward.

The flux across B is the x-component of the field at that face, times the area of the face. The flux across A is similar but it is negative since the flux on that side is into the box. These two faces make a contribution to the total flux given by

$$[U_x(x + dx) - U_x(x)] \, dydz^{27}$$

Note that the quantity in brackets is almost the derivative of U_x with respect to x. It is missing the dx in the 'denominator'. Therefore the flux can be written in terms of the derivative if the numerator and denominator are multiplied by dx:

$$\frac{\partial U_x}{\partial x} \, dxdydz$$

where the partial derivative ∂'s are used to indicate that U_x may vary also with y and z. These changes do not influence the flux because we just moved from face A to B and hence, are looking only at its change with respect to x.

---

[27] Compare this expression to $\frac{df}{dx} = \lim_{\Delta x \to 0} \frac{f(x+\Delta x)-f(x)}{\Delta x}$.

The same goes for the **y**- and **z**-directions. The total flux out of the differential box is the sum of these contributions:

$$\left(\frac{\partial U_x}{\partial x} + \frac{\partial U_y}{\partial y} + \frac{\partial U_z}{\partial z}\right)dxdydz$$

The quantity in parentheses by the way is called the **divergence** of $\vec{U}$. It is a measure of how much $\vec{U}$ is emanating outward from the differential region at that location. It is the flux out of the differential box. (The word **flux** usually refers to finite regions; **divergence** usually refers to infinitesmal regions.) (The divergence of a vector $\vec{F}$ is written $div(\vec{F})$ or $\vec{\nabla} * \vec{F}$.)

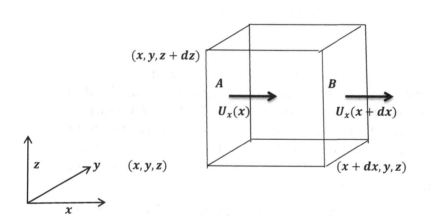

figure 6.14

The Differential Form of Gauss' Law

We have just found the left side of the flux equation. The right side of the equation is $(-4\pi Gm)$. An expression for the mass **m** enclosed by the box is needed. This is just the mass density times the volume. The volume of the differential piece (in Cartesian coördinates) is $dxdydz$; let $\rho(x,y,z)$ represent the mass density (the mass per unit volume, that is) at the point (x,y,z). Then the mass enclosed enclosed by the box is the mass per unit volume where the box is times the volume of the box:

$$\rho(x,y,z)dxdydz\,.$$

Using the two previous expressions in Gauss' law, you get

$$\left(\frac{\partial U_x}{\partial x}+\frac{\partial U_y}{\partial y}+\frac{\partial U_z}{\partial z}\right)dxdydz\ =\ -4\pi G\rho(x,y,z)\,dxdydz$$

$$\left(\frac{\partial U_x}{\partial x}\ +\ \frac{\partial U_y}{\partial y}\ +\ \frac{\partial U_z}{\partial z}\right)\ =\ -4\pi G\rho(x,y,z)$$

$$or\quad div(\vec{U})\ =\ -4\pi G\rho(x,y,z)$$

$$or\quad \vec{\nabla}*\vec{U}\ =\ -4\pi G\rho(x,y,z)$$

The last three equations are the result written in three different ways; all three versions appear in the books. It is this equation that the gravitational field itself must satisfy. Notice that at any point where there is no mass, the right side of the equation is zero. This does not mean that the field is zero; it just means that the local field is the result of mass distributions which are elsewhere.

In general, problems in field theory are more easily approached from the equation for the potential rather than that for the field itself. To write the differential equation for the potential, use

$$U_s\ =\ -\frac{\partial W}{\partial s}\,.$$

where W is the potential function. The equation then, in terms of W is

$$\frac{\partial^2 W}{\partial x^2} + \frac{\partial^2 W}{\partial y^2} + \frac{\partial^2 W}{\partial z^2} = 4\pi G\rho(x,y,z)$$

which is often written

$$\nabla^2 W = 4\pi G\rho(x,y,z)$$

Discussion

The last equation written above is called Poisson's equation. It is used whenever there is a mass (charge) distribution at the points where the field is to be determined. When the source density is zero, the right side of the equation is zero and in this form it is called Laplace's equation:

$$\nabla^2 W = 0.$$

The general solution to LaPlace's equation is known. Its solution has an infinite number of arbitrary constants of integration. The bulk of the work to determine a field is to fit these constants to the conditions of the problem. The infinitude of arbitrary constants is a necessity because it must be possible to fit the single general solution to any one of an infinite number of configurations of source masses (charges).

There are mathematical methods for solving the above equations to find the field configuration that results from a given source distribution. But that is not the point here. What is important in this context is to form an appreciation for the overall view of field problems and the methods for solving them. More specifically this implies you acquire an understanding that:

1. *a field configuration is the result of adding contributions to the field or to its potential from the differential pieces of the source(s);*

495

2. *an inverse square dependence of a field that emanates from a point source is the natural 'thinning out' that results from the geometry of three dimensions;*

3. *any field which is the superposition of inverse square contributions from an arbitrary conglomerate of point sources, discrete or continuous, satisfies Gauss' Law, i.e. the flux of such a field across an arbitrary closed surface is proportional to the total field sources enclosed by the surface;*

4. *for source-configurations that exhibit certain symmetries, Gauss' law can be applied in its integral form to find the field configuration;*

5. *in the general case, the partial differential equations used to find the fields are derived by applying Gauss' law to a closed differential region;*

6. *the flux of the field across a closed differential surface is what is defined as the divergence of the field at that location.*

A familiarity with these things fosters a prowess in handling fields.

END

About the author ---

Mr. Sparapany attended MIT on a partial scholarship where he received his bachelor's degree in physics in 1957. During his twelve years as a member of this community, much of his time was spent teaching physics and mathematics to undergraduates. The consensus of those who were associated with him in this context is that he has a knack for clarifying and simplifying issues often considered to be complex and elusive, imparting to his students an understanding of the subject material that is both deep and foundationally solid. This type of feedback pervades his entire teaching career and, in fact, was a major factor in prompting the writing of this book.

After leaving the MIT community, Mr. Sparapany's endeavors were focused on a number of diverse scientific issues extant at the time. These included the design and construction of special microwave devices for use in communications and radar detection and also the understanding of spoken language by computer. There were two problems addressed during that time which were of particular interest because they required an original cutting edge approach for their mathematical description:

1. The removal of the contamination introduced into telescopic seeing by the strain in the reflecting mirror caused by its own weight;
2. The analysis of entropy-based decision criteria and their application to adaptive pattern recognition and artificial intelligence.

The first of these is one particular 'next step' in the older problem of improving telescopic imagery; the second represents a theory by which a machine (computer) starting with certain abilities can improve them as it operates or, stated a little differently, the theoretical analysis of a process by which a computer learns from its own experience.

But Mr. Sparapany is a jack-of-all-trades --- he also dabbles in painting (see the back cover of this book), he is an accomplished pianist who has performed and taught piano, he composes sacred choral music and has worked both as an accompanist and a church organist.

Additionally, he has written two math books: one is a brief introduction to linear analysis which deals with the algebra of matrices and vectors; the other is a complete solution to Pell's equation together with an exposition of some topics related to it. On a more offbeat side he has written the story for a movie which satirizes the political antics involved in the machinations of the Cold War between Russia and the United States. And finally, he has also written a children's story.

Acknowledgements:

Front Cover: Joseph Mariano
               joemariano393@aol.com

Back Cover: Richard Sparapany
               richpk25@gmail.com

Printed in the United States
By Bookmasters